GEOLOGICAL SURVEY OF CANADA
ECONOMIC GEOLOGY REPORT 36

CANADIAN MINERAL DEPOSIT TYPES: A GEOLOGICAL SYNOPSIS

edited by
O.R. ECKSTRAND

1984

Available in Canada through

authorized bookstore agents and other bookstores

or by mail from

Canadian Government Publishing Centre
Supply and Services Canada
Ottawa, Canada K1A 0S9

and from

Geological Survey of Canada offices:

601 Booth Street
Ottawa, Canada K1A 0E8

3303-33rd Street N.W.,
Calgary, Alberta T2L 2A7

100 West Pender Street
Vancouver, British Columbia V6B 1R8
(mainly B.C. and Yukon)

A deposit copy of this publication is also available
for reference in public libraries across Canada

Cat. No. M43-36/1984E Canada: $8.00
ISBN 0-660-11731-2 Other countries: $9.60

Price subject to change without notice

COVER PHOTOGRAPHS

The cover photographs represent ores that yield six of the ten most valuable mineral commodities presently produced in Canada (copper, gold, iron ore, zinc, silver, asbestos).

Top left to right

Magnetite-jasper iron ore, Sherman Mine, Temagami, Ontario. Mineral deposit type 2.3. (Field of view is about 18 cm)

Massive sulphide ore from the Millenbach copper-zinc mine, consisting mainly of chalcopyrite, with streaks of pyrrhotite and clusters of pyrite metacrysts. Mineral deposit type 9.1. (Length of specimen is 12 cm)

Asbestos ore from the Normandie Mine, Thetford Mines, Quebec, showing 2 to 4 mm wide veins of cross fiber chrysotile asbestos in serpentinized peridotite. Mineral deposit type 26. (Length of specimen is 10.5 cm)

Bottom

Native gold and sphalerite in vein quartz from the Pamour No. 1 Mine, Timmins, Ontario. Mineral deposit type 11. (Length of specimen is 12 cm) Photograph courtesy of Royal Ontario Museum, ROM specimen M41070.

CONTENTS

1	Abstract/Résumé
1	Objective and scope
1	On distinguishing mineral deposit types
2	Names of mineral deposit types
2	Arrangement of mineral deposit types
2	Format of deposit type descriptions
3	Acknowledgments
13	Summaries of Canadian mineral deposit types
69	Selected bibliography
4	Table 1 Canadian Mineral Deposit Types
83	Appendix 1: Locations of Canadian mineral deposits cited in the "Summaries" section.

LISTING OF DEPOSIT TYPES AND AUTHORS

Page	Deposit Types	Authors
13	1 EVAPORITES AND BRINES	R.V. Kirkham
16	2 IRON-RICH SEDIMENTARY STRATA	G.A. Gross
16	2.1 Ironstone	
17	2.2 Iron Formation (Lake Superior Type)	
19	2.3 Iron Formation (Algoma Type)	
20	3 ENRICHED IRON FORMATION	G.A. Gross
21	4 STRATIFORM PHOSPHATE (PHOSPHORITE)	R.L. Christie
22	5 PLACER URANIUM, GOLD	
22	5.1 Pyritic Paleoplacer Uranium, Gold	S.M. Roscoe
23	5.2 Placer Gold	C.R. McLeod
25	6 STRATABOUND SEDIMENT-HOSTED LEAD, ZINC, COPPER, URANIUM	
25	6.1 Mississippi Valley Lead-Zinc	D.F. Sangster
26	6.2 Sandstone Lead	D.F. Sangster
27	6.3 Sedimentary Copper	R.V. Kirkham
28	6.4 Sandstone Uranium	V. Ruzicka, R.T. Bell
29	7 CHEMICAL-SEDIMENT-HOSTED GOLD	R.I. Thorpe, J.M. Franklin
30	8 CLASTIC-SEDIMENT-HOSTED GOLD	R.I. Thorpe
30	8.1 Carbonaceous Shale/Carbonate-Hosted Gold (Carlin Type)	
31	8.2 Turbidite-Hosted Vein and Shear Zone Gold	
33	9 STRATIFORM SULPHIDE, BARITE	
33	9.1 Volcanic-Associated Massive Sulphide	J.M. Franklin, J.W. Lydon, D.F. Sangster
35	9.2 Sediment-Hosted Sulphide	J.W. Lydon, D.F. Sangster
36	9.3 Sediment-Hosted Barite	J.W. Lydon, K.R. Dawson
37	10 VOLCANIC REDBED COPPER	R.V. Kirkham
38	11 VOLCANIC-ASSOCIATED VEIN AND SHEAR ZONE GOLD	R.I. Thorpe, J.M. Franklin
39	12 MAGMATIC NICKEL, COPPER, PLATINUM GROUP ELEMENTS	O.R. Eckstrand
39	12.1 Ultramafic-Associated Nickel-Copper	
41	12.2 Gabbroid-Associated Nickel, Copper, Platinum Group Elements	
43	13 MAFIC/ULTRAMAFIC-HOSTED CHROMITE	J.M. Duke
43	13.1 Stratiform	
45	13.2 Podiform	
46	14 MAFIC INTRUSION-HOSTED TITANIUM-IRON	G.A. Gross, E.R. Rose
47	15 INTRUSION-ASSOCIATED GOLD	R.I. Thorpe, J.M. Franklin
48	16 CARBONATITE-HOSTED DEPOSITS	K.R. Dawson, K.L. Currie
51	17 PORPHYRY COPPER, MOLYBDENUM, TUNGSTEN	R.V. Kirkham, W.D. Sinclair
53	18 VEIN-STOCKWORK TIN	W.D. Sinclair, R.V. Kirkham
55	19 SKARN DEPOSITS	
55	19.1 Skarn Tungsten	K.M. Dawson
56	19.2 Skarn Zinc-Lead-Silver	K.M. Dawson, D.F. Sangster
57	19.3 Skarn Iron	G.A. Gross
58	19.4 Skarn Copper	R.V. Kirkham, W.D. Sinclair
59	20 NEPHELINE- AND CORUNDUM-BEARING ALKALINE GNEISSES	L. Moyd
61	21 UNCONFORMITY-ASSOCIATED URANIUM	L.P. Tremblay, V. Ruzicka
63	22 ARSENIDE VEIN SILVER, URANIUM	R.I. Thorpe
64	23 VEIN URANIUM	L.P. Tremblay, V. Ruzicka
65	24 VEIN COPPER	R.V. Kirkham
66	25 FELSIC INTRUSION-ASSOCIATED SILVER-LEAD-ZINC VEINS	D.F. Sangster
67	26 ULTRAMAFIC-HOSTED ASBESTOS	J.M. Duke

CANADIAN MINERAL DEPOSIT TYPES: A GEOLOGICAL SYNOPSIS

Abstract

Summaries of the main geological characteristics of 40 Canadian mineral deposit types are presented in a systematic format. About one half of these deposit types account for the bulk of Canadian metal and industrial mineral production. The definition of a mineral deposit type is based on recognition of a group of deposits whose geological features are sufficiently similar that they suggest a common genesis, whether or not the genesis is well understood. The summaries list contained commodities and names of example deposits, and outline economic significance, typical size and grade of orebodies, geological setting, host rocks, associated rocks, form and distribution of mineralization, minerals present, age, genetic models, and guides to exploration. A selected bibliography provides an introduction to the relevant literature.

Résumé

Le présent ouvrage offre une présentation systématique des résumés des principales caractéristiques géologiques de 40 types de gisements minéraux du Canada. Environ la moitié de ces gisements types représentent d'ailleurs la majorité de la production canadienne de métaux et de minéraux industriels. La définition d'un type de gisement minéral se fonde sur l'identification d'un groupe de gisements dont les caractéristiques géologiques s'apparentent au point où elles semblent indiquer une origine commune, peu importe qu'il s'agisse ou non d'une origine bien connue. Les résumés indiquent les produits contenus et le nom de gisements modèles, et donne un aperçu de l'importance économique, des dimensions caractéristiques et de la qualité des masses de minerai, du cadre géologique, des roches minéralisées, des roches connexes, de la forme et de la répartition des zones minéralisées, des minéraux présents, de leur âge, des modèles liés à la génèse et des guides conçus aux fins d'exploration. La bibliographie sert d'introduction à des ouvrages utiles.

Objective and Scope

The objective of this report is to define and characterize in summary form the most important types of Canadian mineral deposits. The resulting collection of summaries constitutes a "synopsis" of Canadian mineral deposit types.

Recognition of "types" of mineral deposits is a convenience practiced widely by economic geologists in both mineral exploration and research on ore genesis. In mineral exploration, geologists base their exploration programs on the characteristics of the particular type of mineral deposit that they hope to discover. Geologists engaged in research on mineral deposits tend to regard the known characteristics of a deposit type as a summary of the main facts that any genetic hypothesis must explain. Hence the concept of mineral deposit types is of considerable importance to economic geologists.

This synopsis was prepared by economic geologists of the Geological Survey of Canada as a byproduct of their continuing studies of Canadian mineral deposits. As it reflects the current general understanding of Canadian mineral deposits and because the definitions of deposit types correspond closely to those in common use, this synopsis may be useful to geologists and others involved in the mining and mineral industry in Canada.

The types of mineral deposits treated in this report are those of metallic minerals and some industrial minerals, but those of fossil fuels are not included. Only the deposit types that have some significance in Canada are considered. A deposit type was included if it accounted for at least a modest or historically significant amount of Canadian production or reserves, or if it was represented by mineral occurrences and was judged to have potential for significant undiscovered deposits in Canada. These obviously flexible guidelines have led to some arbitrary choices of the deposit types to be included.

Exclusions of some deposit types have been mainly for one or more of the following reasons:

1. uncertain affiliations of some deposits (in some cases because of superimposed metamorphism and/or deformation);
2. economic insignificance (in Canada); and
3. lack of accessible information.

This synopsis does not constitute a true classification of mineral deposits. A classification implies completeness of coverage, i.e., a pigeon hole for every known deposit. A classification also implies that it has been derived by systematic application of certain principles. The synopsis, however, fulfills neither of these criteria; it is simply an enumeration of more or less distinct types, each made up of recognizably similar mineral deposits. It would be more correct to view the synopsis as a potential basis for a classification of mineral deposits.

This report presents:

1. a list of the main Canadian mineral deposit types;
2. a highly condensed summary for each deposit type of its geological and economic characteristics, genetic interpretation, and guides to exploration;
3. a selected bibliography for each type; and
4. photographs illustrating selected features of some deposit types.

On distinguishing mineral deposit types

For purposes of this report, a mineral deposit type is defined as a hypothetical composite of the geological characteristics common to a group of similar mineral deposits. It is implicit that such deposits, because of their similarities, are expected to have a common mode of genesis, whether or not that mode of genesis is well understood.

The principal means of identifying a mineral deposit type is to note that several mineral deposits appear to have similar characteristics, such as host rocks, commodities, geological environments, form of deposit, and mineralogy. Many of the mineral deposit types cited here have been recognized for a long time; for example, veins of gold and copper and the ironstone type of sedimentary iron deposits. Others such as "volcanic redbed copper deposits" have not previously been recognized as deposit types.

Recognition of various mineral deposit types evolves continuously as a result of discovery and study of new mineral deposits, and of advances in research on ore-forming processes and the paleoenvironments in which mineral deposits have formed. One natural consequence of such evolution is the recognition of hierarchical relationships amongst deposit types. For example, some deposit types seem better regarded as variants or subtypes of a single type rather than as distinct types: placer gold deposits of the Klondike and paleoplacer uranium deposits of Elliot Lake are a case in point. In this synopsis three levels of hierarchy have been used, though in some cases choice of level was rather arbitrary and may be subject to revision.

An inherent difficulty in devising a uniform treatment of deposit types is the uneven state of our understanding of mineral deposits. For instance, the many studies on volcanic-associated massive sulphide deposits, porphyry copper deposits and ultramafic-associated nickel deposits have led to a modest understanding of their genesis, and has resulted in a relatively clear definition of their types and subtypes. Many other deposits (e.g. vein deposits of many commodities), however, have not been studied in detail, and recognition of their types is correspondingly vague. It seems probable, therefore, that some of these types have been defined in either too narrow or too broad a manner.

The condensed format of this synopsis does not permit elaboration of the relationships amongst deposit types. Because of this and other previously mentioned constraints, this synopsis must be regarded as a preliminary and tentative definition of deposit types.

Names of mineral deposit types

The scheme adopted for the naming of deposit types uses a mixture of traditional and newly proposed terms. Traditional terms such as "placer gold" and "Mississippi Valley lead-zinc" have been retained. The new terms are intended to be informative, and in most cases consist of two parts. The first part refers to the most noteworthy geological characteristic of that type, which in many cases is the host rock or associated rock. Thus "sandstone-hosted" refers to a type that occurs in sandstone, and "intrusion-associated" refers to one that occurs in or near intrusions. The terms "-hosted" and "-associated" are intended to indicate only a spatial relationship between rock and ore, although in many instances a genetic relationship is also accepted. The second part of the name generally identifies the main commodities contained in the deposit type. Although this two-part system of nomenclature results in some cumbersome names, they are considered justified for the sake of clarity.

Arrangement of mineral deposit types

The list of mineral deposit types (see Table 1) is loosely organized according to the following scheme:

Affiliation	Mineral Deposit Types
Sedimentary	1 to 9
Volcanic	9 to 12
Intrusive	12 to 19
Hydrothermal-metasomatic	15 to 26
Structural	21 to 26 and others.

This scheme is but one of several that could have been used. The overlap between adjacent affiliations in this scheme indicates only a few of the multiple affiliations that some deposit types show. For instance, in the present treatment sandstone uranium deposits (6.4) and unconformity-associated deposits (21) appear in distinct, apparently unrelated affiliations; some geologists, however, consider them to be genetically related and would place the two types in the same group. Consequently, because of its subjective nature, the above scheme has not been used formally in Table 1 for grouping deposit types.

Format

To facilitate concise descriptions of deposit types, the adopted format employs a set of standardized headings, as well as certain conventions for presentation of the data. The headings are arranged in groups, according to the following scheme:

Purpose of the information	Heading
Identification of deposit type	- Name of type, subtype, secondary subtype - Commodities - Examples
Economic characteristics	- Importance - Typical grade, tonnage
Geological characteristics	- Geological setting - Host rocks or mineralized rocks - Associated rocks - Form of deposit, distribution of ore minerals - Minerals - Age, host rocks - Age, ore
Interpretation, application	- Genetic model - Ore controls, guides to exploration

A brief selected bibliography for each type is given at the end of the report.

The following comments for each of the headings describe the nature, scope, organization and limitations of the contained information.

Name of deposit type, subtype, secondary subtypes: Each deposit type (e.g. 5) or subtype (e.g. 5.1) is accorded a separate summary (about one page). Secondary subtypes (e.g. 5.1.a, 5.1.b) are treated as variants of the type or subtype, and are singled out only under those headings in which they show distinctive characteristics. In the names, commodities linked by hyphens (e.g. Mississippi Valley Lead-Zinc) consistently occur together; those separated by commas (e.g. Placer Uranium, Gold) do not occur together in all deposits.

Commodities: The commodities listed first constitute the principal recovered products at one or more deposits of the type in question. However, all the commodities so listed are not necessarily recovered from all deposits of that type. Commodities enclosed within parentheses include byproduct commodities and chemical elements that tend to be characteristic or diagnostic of the deposit type in question.

Examples: Authors used considerable latitude in their choice of examples, and treatment is therefore variable. Some authors listed only a few of the best examples, whereas others provided extensive lists. Some confined their choices to Canadian examples, but others cited numerous, generally classic, foreign examples as well, especially where good Canadian examples are few.

Importance: This section gives some indication of the economic significance of the deposit type in question in a Canadian and/or global context. The indication may be in terms of past or current production, or of reserves or anticipated potential.

Typical grade, tonnage: This section is intended to provide an appreciation of typical size and grade of deposits of the type in question. Where possible, grade and tonnage of individual ore bodies are given, but in some cases the available data for deposits include unspecified numbers of ore bodies. In other cases the ore zones are not readily separable into distinct bodies but are linked, and do not lend themselves to simple statements of size; in these cases the figures cited may be for mines or whole districts, as stated in the individual descriptions.

Geological setting: This section indicates the broad geological setting of the rocks in which most deposits occur. In some cases, however, the broad scale setting is poorly known, and the description given is for the more local setting.

Host rocks or mineralized rocks: The lithologic character of the ore is described herein. This may refer to the host rock in instances where ore minerals are dispersed through the rock, or to the "ore-rock" in cases where the ore minerals constitute a large part of the rock (e.g. salt and gypsum beds, iron formation).

Associated rocks: This section presents information on the lithologic character of rocks that enclose, adjoin, or occur spatially near the host rocks or mineralized rocks, especially those rocks that may have some genetic association with the deposits. In two cases ("porphyry copper, molybdenum, tungsten", and "skarn iron") this section has been combined with "Host rocks or mineralized rocks" because the classification as host rock or associated rock is rather arbitrary.

Form of deposit, distribution of ore minerals: This section describes the typical geometric shape of ore bodies, their physical and structural relationship to wall rocks and associated rocks, and the distribution of ore minerals (e.g., disseminations, veins, masses). If present, characteristic zonation of ore minerals and alteration is also described.

Minerals: The principal ore minerals from which the main commodities are derived are listed first. Other associated minerals follow in brackets, and include less abundant ore minerals and typical gangue minerals, especially those that tend to be diagnostic of the deposit type.

Age, host rocks: The ages of host rocks are given either as a generalization or individually for specific example deposits. They are in the form of either absolute dates, or geological eons, eras or periods.

Age, ore: The age of ore emplacement is given either in the same manner as for host rocks, or more commonly, in terms relative to the host rocks or a tectonic event in the area (e.g., penecontemporaneous with host sedimentary rocks).

Genetic model: An attempt has been made to present objectively the currently accepted models, and to avoid the more contentious ones. Undoubtedly, however, some of the authors have leaned toward their preferred theories. For poorly understood deposit types, the suggested models tend to be speculative and/or vague.

Ore controls, guides to exploration: The geological guides to exploration are based on both genetic models and empirically observed geological relationships. They are oriented mainly toward features that may be useful for discovery of deposits, but in some cases the guides point out factors that affect the economic value of deposits; for example, the beneficial influence on sedimentary iron ores of repetition of beds by folding and/or faulting. Geophysical methods receive mention only in a few special cases, for instance, the recognition of buried carbonatite intrusions by their distinctive annular magnetic anomalies.

Authors: Most of the authors are members of the Economic Geology Division, Geological Survey of Canada. Others are identified as to organizational affiliation.

Selected Bibliography: References for each deposit type or subtype are given at the end of the publication. The references listed are intended only as an introduction to relevant literature, and average about five references for each type or subtype. Preference has been given to recent review papers or symposia, where available, that deal broadly with the mineral deposit type in question. Failing this, references to descriptions of typical example deposits are given.

Acknowledgments

Many authors have contributed constructive comments on deposit type summaries other than their own. Exceptional contributions of this nature as well as overall guidance in determining the scope and structure of this synopsis require the special recognition of R.V. Kirkham, S.M. Roscoe, and R.I. Thorpe. Finally, I gratefully acknowledge R.V. Kirkham's continuous, enthusiastic encouragement and support, and B. Williamson's care and assistance in preparation of the manuscript.

TABLE 1

CANADIAN MINERAL DEPOSIT TYPES

1 **EVAPORITES AND BRINES**
* 1.a Marine
1.b Nonmarine

2 **IRON-RICH SEDIMENTARY STRATA**
2.1 **Ironstone**
* 2.2 **Iron Formation (Lake Superior Type)**
* 2.3 **Iron Formation (Algoma Type)**

* 3 **ENRICHED IRON FORMATION**

4 **STRATIFORM PHOSPHATE (PHOSPHORITE)**
4.a Miogeosynclinal
4.b Platformal

5 **PLACER URANIUM, GOLD**
5.1 **Pyritic Paleoplacer Uranium, Gold**
* 5.1.a Uranium
5.1.b Gold
* 5.2 **Placer Gold**

6 **STRATABOUND SEDIMENT-HOSTED LEAD, ZINC, COPPER, URANIUM**
* 6.1 **Mississippi Valley Lead-Zinc**
6.2 **Sandstone Lead**
6.3 **Sedimentary Copper**
6.3.a Paralic Marine (Kupferschiefer Type)
6.3.b Continental (Red Bed Type)
6.4 **Sandstone Uranium**

7 **CHEMICAL-SEDIMENT-HOSTED GOLD**
7.a Carbonate-Oxide Iron Formation
7.b Arsenical Sulphide-Silicate Iron Formation
7.c Stratiform Pyrite
7.d Chert-Sulphide

8 **CLASTIC-SEDIMENT-HOSTED GOLD**
8.1 **Carbonaceous Shale/Carbonate-Hosted Gold (Carlin Type)**
8.2 **Turbidite-Hosted Vein and Shear Zone Gold**

9 **STRATIFORM SULPHIDE, BARITE**
* 9.1 **Volcanic-associated Massive Sulphide**
9.1.a Copper-Zinc
9.1.b Zinc-Lead-Copper
* 9.2 **Sediment-Hosted Sulphide**
9.3 **Sediment-Hosted Barite**

10 **VOLCANIC RED BED COPPER**

*11 **VOLCANIC-ASSOCIATED VEIN AND SHEAR ZONE GOLD**

12 **MAGMATIC NICKEL, COPPER, PLATINUM GROUP ELEMENTS**
12.1 **Ultramafic-associated Nickel-Copper**
12.1.a Volcanic Peridotite Nickel
* 12.1.b Intrusive Dunite Nickel
12.1.c Intrusive Ultramafic Nickel-Copper
12.2 **Gabbroid-associated Nickel, Copper, Platinum Group Elements**
* 12.2.a Layered Intrusive, Nickel-Copper
12.2.b Layered Intrusive, Platinum Group Elements
12.2.c Stock

13 **MAFIC/ULTRAMAFIC-HOSTED CHROMITE**
13.1 **Stratiform**
13.2 **Podiform**

14 **MAFIC INTRUSION-HOSTED TITANIUM-IRON**
* 14.a Anorthosite-Hosted Ilmenite
14.b Gabbroic-Anorthosite-Hosted Titaniferous Magnetite

15 **INTRUSION-ASSOCIATED GOLD**
* 15.a Sub-alkalic Felsic
15.b Alkalic
15.c Mafic

16 **CARBONATITE-HOSTED DEPOSITS**
16.a Nephelinitic Carbonatite
16.b Ultramafic Carbonatite

17 **PORPHYRY COPPER, MOLYBDENUM, TUNGSTEN**
* 17.a Calc-alkalic – associated Copper, Molybdenum
* 17.b Alkalic – associated Copper
* 17.c Calc-alkalic – associated Molybdenum, Tungsten

18 **VEIN-STOCKWORK TIN**

19 **SKARN DEPOSITS**
* 19.1 **Skarn Tungsten**
19.2 **Skarn Zinc-Lead-Silver**
19.3 **Skarn Iron**
19.3.a Intrusion-associated (Contact Metasomatic)
19.3.b Stratiform in Metamorphic Terrane
* 19.4 **Skarn Copper**

20 **NEPHELINE-AND CORUNDUM-BEARING ALKALINE GNEISSES**
20.a "Nepheline Syenite"
20.b Corundum
20.c Molybdenum

*21 **UNCONFORMITY-ASSOCIATED URANIUM**

*22 **ARSENIDE VEIN SILVER, URANIUM**

23 **VEIN URANIUM**

24 **VEIN COPPER**

25 **FELSIC INTRUSION-ASSOCIATED SILVER-LEAD-ZINC VEINS**

*26 **ULTRAMAFIC-HOSTED ASBESTOS**

* Deposit types that have contributed to a significant proportion of past or present Canadian mineral production.

Plate 1. *2.2 Iron Formation (Lake Superior type). Near Schefferville, Quebec-Labrador iron ore belt. Jasper-hematite-magnetite facies of iron formation, showing typical, relatively thin, well preserved sedimentary macro-bedding. Magnet is 3.5 cm. Photo: G.A. Gross.*

Plate 2. *6.1 Mississippi Valley Lead-Zinc. Pine Point, Northwest Territories. Typical botryoidal textured sulphides. Layered sphalerite (dark brown and white) overlain by coarse grained galena (steel blue). Sulphides precipitated in cavity beginning at lower left, with successive layers added into open space to right; last sulphide deposited was galena. Photo: D.F. Sangster.*

Plate 3. *3 Enriched Iron Formation. French mine, Schefferville area, Quebec. Red (R), yellow (Y), and blue (B) types of hematite-goethite iron ore derived from slate, silicate-carbonate, and jasper oxide facies respectively of Lake Superior type iron formation. Note the transition from leached red ferruginous slate (S) to red ore. (Bench height about 10 m). Photo: G.A. Gross.*

Plate 4. *5.1 Pyritic Paleoplacer Uranium, Gold. Elliot lake, Ontario. Uraniferous, pyritic quartz-pebble conglomerate from the AB reef, Denison mine. The uranium minerals (mainly uraninite and a uranium-titanium phase; see Pl. 5) occur in the matrix, associated with pyrite (white) and other heavy minerals. Field view is 14 cm wide. Sample: V. Ruzicka.*

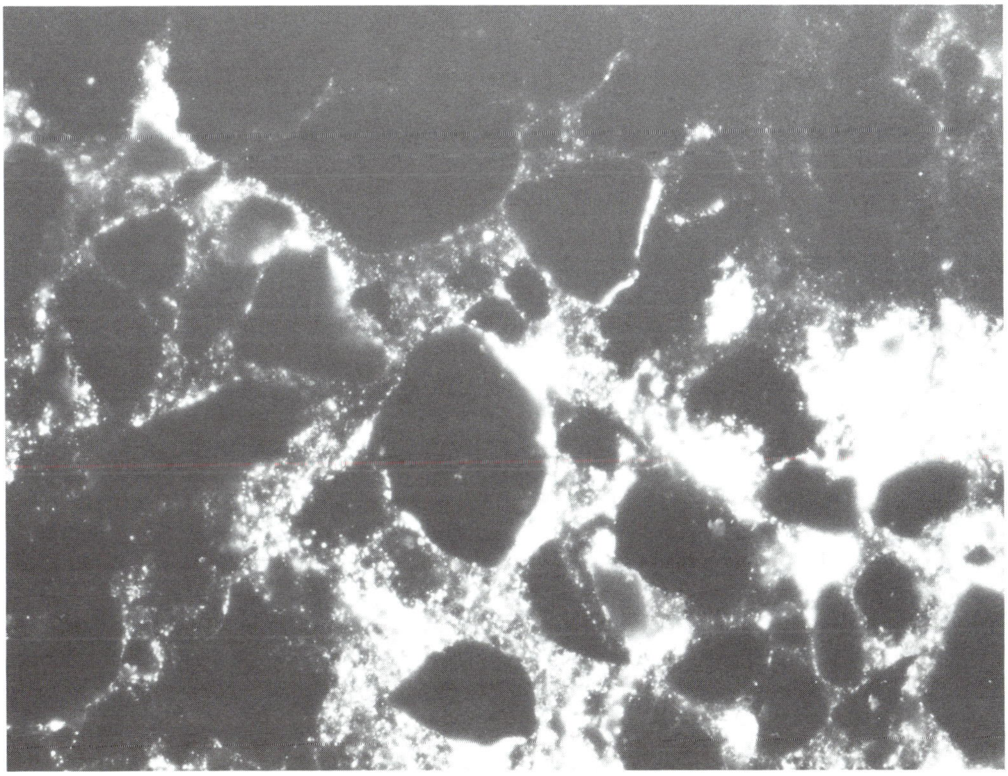

Plate 5. *5.1 Pyritic Paleoplacer Uranium, Gold. Elliot Lake, Ontario. Autoradiograph of the sample shown in Plate 4. The white areas are caused by radiation from uranium- and thorium-bearing minerals. The distribution of radioactive minerals and pyrite (cf. Pl. 4) is similar, but different in detail. Autoradiograph contributed by V. Ruzicka.*

Plate 6. *9.1 Volcanic-Associated Massive Sulphide. Lake Dufault mine, Noranda, Quebec. Typical well bedded, massive ore: pyrite (dull brassy yellow), chalcopyrite (bright yellow), sphalerite (dark metallic blue-grey). Photo: R.V. Kirkham.*

Plate 7. *9.1 Volcanic-Associated Massive Sulphide. MacLean deposit, Buchans mine, Newfoundland. Typical fragmental sulphide ore: sphalerite (dark metallic grey), galena (light metallic grey), chalcopyrite (bright yellow), pyrite (dull brassy yellow). Most of the lithic clasts are altered volcanic rocks. Sample is about 15 cm long. Photo: R.V. Kirkham.*

Plate 8. *9.1 Volcanic-Associated Massive Sulphide. York Harbour deposit, Newfoundland. Typical fragmental massive sphalerite (grey)-chalcopyrite-pyrite ore. Note contorted, well laminated chalcopyrite clast. Photo: R.V. Kirkham.*

Plate 9. *9.2 Sediment-Hosted Sulphide. Sullivan mine, Kimberley, British Columbia. Highly contorted layered sulphides (pyrrhotite, sphalerite, galena) and volcaniclastic sediments, Main Band, Sullivan deposit. Lighter colored layers are volcaniclastic sediments; darker layers are sulphides darkened by oxidation; silver coloured patch left of hammer is freshly exposed sulphides. Photo: D.F. Sangster.*

Plate 10. *9.3 Sediment-Hosted Barite. TEA property, Macmillan Pass area, Yukon. Interbedded barite (ba) and limestone (ls). Note finely laminated internal structure of both barite and limestone beds. Height of field of view is about 80 cm. Photo: J.W. Lydon.*

Plate 11. *10 Volcanic Red Bed Copper. Sustut deposit, British Columbia. Typical, finely disseminated native copper (bright yellow specks) in maroon Upper Triassic lapilli tuff of the Takla Group. Width of field of view is about 3.5 cm. Photo: R.V. Kirkham.*

Plate 12. *14.b Mafic Intrusion-Hosted Titanium-Iron. Chibougamau, Quebec. Vanadiferous magnetite-ilmenite layers interstratified with anorthositic metagabbro and metapyroxenite layers of the Doré Lake Complex. Photo: O.R. Eckstrand.*

Plate 13. *19.4 Skarn Copper. Craigmont mine, British Columbia. Coarsely crystalline copper ore showing potash feldspar (pink), magnetite (grey), chalcopyrite (bright yellow), calcite (white), chlorite (black). Photo: R.V. Kirkham.*

Plate 14. *23 Vein Uranium. Kazan Falls, Baker Lake area, Northwest Territories. Red hematitic staining accompanies fracture fillings of pitchblende (not visible) along a mylonitic fault zone traversing layered gneiss. Photo: A.R. Miller.*

Plate 15. *26 Ultramafic-Hosted Asbestos. Thetford Mines, Quebec. Cross-fibre asbestos ore. Note the generally uniform spacing of the chrysotile veins in serpentinized dunite and the en echelon alignment of wall rock inclusions in the veins. Sample: M. Larose. Specimen is about 10.5 cm long.*

1. EVAPORITES AND BRINES

1.a Marine
1.b Nonmarine

COMMODITIES

(1.a) NaCl, KCl, gypsum (S, Mg, Sr, Cl, Br, I)
(1.b) Trona, borates, Li, NaCl, Na_2SO_4 (KCl, Mg, Br, P, W, nitrates, zeolites, U, I, clay minerals)

EXAMPLES:
Canadian – ***Foreign***

(1.a) Salina Formation, Ont. (NaCl, gypsum); Windsor Group, N.S.-N.B.-Nfld.-Que. (NaCl, KCl, gypsum, celestite); Prairie Formation, Sask. (KCl, NaCl); Gypsumville, Man. (gypsum); Windermere, B.C. (gypsum) – *Michigan, Delaware and Paradox basins, U.S.A. (KCl, NaCl, gypsum); Gulf Coast salt domes, U.S.A. and Mexico (S, NaCl); Zechstein and Messinian basins in Europe, the Ural Mountains in the U.S.S.R., and the Khorat Plateau in Thailand (KCl, NaCl, gypsum)*
(1.b) Modern alkaline lakes, Sask. and Alta. (Na_2SO_4) – *Green River Formation, Wyoming (trona); Searles Lake, California (borates, NaCl, KCl, Mg, Br, I, Li, W); Clayton Valley, Nevada (Li); Great Salt Lake, Utah (NaCl, KCl, Mg, Br); Dead Sea, Israel and Jordan (NaCl, KCl, Mg, Br); Bigadiz, Emet and Kirka, Turkey (borates); Lake Magadi, Kenya (trona); Atacama desert, Chile (nitrates); Basin and Range Province playa lake sediments, western U.S.A. (zeolites); Yeelirrie, Western Australia (U)*

IMPORTANCE

(1.a) Most important source of salt, potash and gypsum in Canada and rest of world. Potential sites for nuclear waste and petroleum storage.
(1.b) Canada: main source of sodium sulphate. Otherwise of minor importance.
World: Important source of trona, borates and lithium in the United States. Searles Lake is the largest single resource of tungsten in the United States, containing 50 to 60% of known reserves. Lacustrine evaporites could become a major source of zeolites. Nitrate deposits in Chile were important for 100 years but are of minor importance now. Calcrete and lacustrine deposits might become a minor source of uranium in parts of the world.

TYPICAL GRADE, TONNAGE

(1.a) 90 to 100% NaCl; 15 to 30% K_2O; 90 to 100% gypsum; tens of millions to greater than one billion tonnes.
Saskatchewan deposits estimated to contain greater than 14 billion tonnes of recoverable K_2O-equivalent (by conventional mining) at an average grade of 25% K_2O; New Brunswick, several hundred million tonnes of ore averaging 20 to 30% K_2O; in Gulf Coast salt domes, individual deposits may contain greater than 50 million tonnes of extractable sulphur.
(1.b) Saskatchewan alkaline lakes are estimated to contain total reserves of 55 to 65 million tonnes of sodium sulphate. Wyoming deposits probably contain greater than 30 billion tonnes of trona; Searles Lake deposits and brines contain greater than 200 million tonnes of extractable soluble salts. Emet area in Turkey was estimated to contain 45 million tonnes averaging 45% B_2O_3 and Kirka 500 million tonnes averaging 27% B_2O_3. Chilean nitrate reserves are greater than 2.5 billion tonnes of ore at a cutoff grade of 7% $NaNO_3$; Yeelirrie is estimated to have 41 000 tonnes contained U_3O_8 at an average grade of 0.15% U_3O_8.

GEOLOGICAL SETTING

(1.a) Large restricted marginal marine basins and extensive coastal sabkhas in arid areas.
(1.b) Large desert salt lakes and inland sabkhas (playas) in arid and semi-arid areas. Many desert sedimentary basins with closed drainage systems are the results of rifting.

HOST ROCKS OR MINERALIZED ROCKS

(1.a) Salt, potash, gypsum and anhydrite beds.
(1.b) Trona, borate, salt, potash, gypsum, glauberite and shale beds; calcretes (caliche); tuffaceous rocks (zeolites); and intrastratal or lacustrine brines.

ASSOCIATED ROCKS

(1.a) Fetid marine shale and limestone, reefal limestones, red beds. Stromatolites and algal mats are common.
(1.b) Oil shale, siltstone, sandstone, conglomerate and red beds. Plant and fish fossils common.

FORM OF DEPOSIT, DISTRIBUTION OF ORE MINERALS

The ores may consist of sedimentary beds; caprocks of salt domes; or surface and subsurface brines. To be economic, the main commodity normally represents a high precentage of the ore unit. Intraformational folding of potash and halite beds is common.

MINERALS: Principal ore minerals
– ***Associated minerals***

(1.a) Gypsum, anhydrite, halite, sylvite, carnallite, polyhalite, tachyhydrite, langbeinite, kainite, celestite, strontianite.
– *Dolomite, calcite, clay minerals, quartz, pyrite*
(1.b) Highly variable; Na and Mg minerals are characteristic. Gypsum, anhydrite, halite, glauberite, trona, nahcolite, borax, kernite, hanksite, burkeite, ulexite, colemanite, mirabilite.
– *Dolomite, calcite, magnesite, siderite, clay minerals, quartz, talc, pyrite, numerous minerals characteristic of evaporite deposits*
Chilean nitrate deposits: Soda niter, niter, darapskite, humberstonite, lautarite, bruggenite, dietzeite. – *Halite, gypsum, anhydrite, glauberite, bloedite, tarapacaite, lopezite, ulexite*
Yeelirrie uranium deposit: Carnotite. – *Calcite, chalcedony, dolomite, gypsum, sepiolite*

AGE, HOST ROCKS

Late Proterozoic to Recent

AGE, ORE

Mostly syngenetic (same age as host rocks) with extensive diagenetic modification. The celestite deposits at Lake Enon, Nova Scotia are thought to be diagenetic. Native sulphur deposits in caprocks of salt domes are thought to have formed epigenetically by bacterial reduction of sulphate.

GENETIC MODEL

Deposition of soluble components is caused generally by evaporation in salinas (salt lakes) and sabkhas (low-lying salt flats) and by precipitation from subsurface brines in both marginal marine and inland desert basins. Many of the economic products such as halite, potassium salts, sodium sulphate, trona, borates, nitrates and carnotite are thought to be syngenetic. At Lake Enon, Nova Scotia, strontium might have been concentrated in an evaporitic environment during diagenesis. Abundant volcanic ash beds in salt lakes are considered favourable for the formation of zeolites. Nitrates are thought to be concentrated only under extremely arid conditions. Calcrete uranium deposits probably require separate sources of uranium and vanadium in large catchment areas, and an evaporitic calcrete environment for formation. Economic gypsum normally results from near-surface hydration of anhydrite.

ORE CONTROLS, GUIDES TO EXPLORATION

1. Evaporitic basins, preferably of large size, or smaller basins with unusual concentrations of elements (e.g. borates, lithium, uranium, nitrates).
2. Identification and outline of distribution (e.g. facies analysis) of valuable components within the evaporite sequence.
3. Mining sites are generally near surface (e.g. open pit gypsum mines); salt domes or anticlines bring salt and potash beds close to surface.
4. Geochemistry of spring waters (Na^+, K^+, Br^+, Cl^-, etc.) has proved useful in discovery of the New Brunswick and Khorat Plateau potash deposits.
5. Gravity and seismic surveys are useful for outlining large salt domes and anticlines.

AUTHOR

R.V. Kirkham

Figure 1. *1.a Evaporites and Brines. Sifto Salt Company Mine, Goderich, Ontario. Relatively undeformed bedded halite, A-2 Unit of the Upper Silurian Salina Formation. Scale indicated by vehicle in foreground. Photo courtesy of Sifto Salt Division, DOMTAR Ltd. (GSC 203368).*

Figure 2. *1.a Evaporites and Brines. Canadian Rock Salt Company Ltd. Mine, Pugwash, Nova Scotia. Highly contorted anhydrite (darker grey) interlayered with bedded halite. Lower Mississippian Windsor Group. Photo: H. Wiele, Bedford Institute of Oceanography (B.I.O. 4392-12).*

2. IRON—RICH SEDIMENTARY STRATA

2.1 IRONSTONE

COMMODITIES
Fe

EXAMPLES: Canadian – *Foreign*
Wabana, Nfld.; Peace River, Alta. – *Clinton Formation, Alabama to New York; Minette ores, England, France, Luxembourg and Germany*

IMPORTANCE
Canada: Mined at Wabana for 75 years. Potential source of iron in Peace River area. Much less important than the iron formation subtypes (2.2, 2.3).
World: The principal domestic source of iron in north-central Europe. Much less important than iron formation subtypes (2.2, 2.3).

TYPICAL GRADE, TONNAGE
Up to billions of tonnes at grades ranging from 30 to 55% Fe, averaging 30 to 35% Fe.

GEOLOGICAL SETTING
In shallow shelf and estuarine sedimentary sequences; in neritic, oxygenated to euxinic and restricted environments.

HOST ROCKS OR MINERALIZED ROCKS
Thin, massive hematite-siderite-chamosite-clay beds with oolitic textures; intercalated with shale and sandstone. Fossil debris common.

ASSOCIATED ROCKS
Black shale, ferruginous sandstone, pyritic shale, limestone, manganiferous and phosphatic shale and sandstone, glauconitic sand.

FORM OF DEPOSIT, DISTRIBUTION OF ORE MINERALS
Thin, massive ore beds commonly 1 to 10 m thick, interbedded with shale and sandstone. Commonly mined only where beds are little disturbed, and have shallow dips.

MINERALS: Principal ore minerals – *Associated minerals*
Hematite, goethite, siderite, chamosite.
– *Calcite, ankerite, various clay minerals, quartz (grains), pyrite, phosphatic fossil debris*

AGE, HOST ROCKS
Precambrian to Recent, most common in lower Paleozoic and Mesozoic sequences.

AGE, ORE
Syngenetic; same age as host rocks.

GENETIC MODEL
Deposition of iron and clay minerals in neritic basins, lagoons and estuaries under conditions ranging from oxygenated to euxinic. Several models are proposed for the formation of deposits of this type:
1. clastic and chemical sediments derived from landmass (the most commonly accepted model);
2. Fe leached from deep, euxinic muds, transported by upwelling currents, and deposited in oxygenated shore basins;
3. hydrothermal effusive derivation from volcanogenic sources;
4. diagenetic replacement of carbonate-rich sediments by iron derived from associated carbonaceous muds and other sediments.

ORE CONTROLS, GUIDES TO EXPLORATION
1. Well defined stratigraphic control.
2. The iron ore beds seem to constitute neritic and estuarine chemical/clastic sedimentary facies characterized by distinctive mineral assemblages and textures.
3. Clean iron ore beds, free from excessive interbedded shale and sandstone are desirable.
4. Excessive structural disruption, steep dips adversely affect mining feasibility.

AUTHOR
G.A. Gross

2.2 IRON FORMATION (LAKE SUPERIOR TYPE)

COMMODITIES	Fe (Mn)
EXAMPLES: Canadian – ***Foreign***	Knob Lake, Wabush Lake and Mount Wright areas, Que. and Lab. – *Mesabi Range, Minnesota; Marquette Range, Michigan; Minas Gerais area, Brazil*
IMPORTANCE	Canada: the major source of iron. World: the major source of iron.
TYPICAL GRADE, TONNAGE	Up to billions of tonnes, at grades ranging from 15 to 45% Fe, averaging 30% Fe.
GEOLOGICAL SETTING	Continental shelves and slopes possibly contemporaneous with offshore volcanic ridges. Principal development in middle Precambrian shelf sequences marginal to Archean cratons.
HOST ROCKS OR MINERALIZED ROCKS	Iron formations consist mainly of iron- and silica-rich beds; common varieties are taconite, itabirite, banded hematite quartzite, and jaspilite; composed of oxide, silicate and carbonate facies, and may also include sulphide facies. Commonly intercalated with other shelf sediments: black carbon-rich shale, red shale, other shale and argillite, tuff, greywacke, quartzite, dolomite.
ASSOCIATED ROCKS	Bedded chert and chert breccia, dolomite, stromatolitic dolomite and chert, black shale, argillite, siltstone, quartzite, conglomerate, redbeds, tuff, lava, volcaniclastic rocks; metamorphic equivalents.
FORM OF DEPOSIT, DISTRIBUTION OF ORE MINERALS	Mineable deposits are sedimentary beds with cumulative thickness typically from 30 to 150 m, and strike length of several kilometres. In many deposits, repetition of beds caused by isoclinal folding or thrust faulting has produced widths that are economically mineable. Ore mineral distribution is largely determined by primary sedimentary deposition. Granular and oolitic textures common.
MINERALS: Principal ore minerals – ***Associated minerals***	Magnetite, hematite, goethite; pyrolusite, manganite, hollandite. – *Finely laminated chert, quartz, Fe-silicates, Fe-carbonates and Fe-sulphides; primary or metamorphic derivatives*
AGE, HOST ROCKS	Precambrian, predominantly early Proterozoic (2.4 to 1.9 Ga).
AGE, ORE	Syngenetic, same age as host rocks. In Canada, major deformation during Hudsonian and, in places, Grenvillian orogenies produced mineable thicknesses of iron formation.
GENETIC MODEL	A preferred model invokes chemical, collodial, and possibly biochemical precipitates of iron and silica in euxinic to oxidizing environments, derived from hydrothermal effusive sources related to fracture systems and offshore volcanic activity. Deposition may be distal from effusive centres and hot spring activity. Other models derive silica and iron from deeply weathered land masses, or by leaching from euxinic sediments. Sedimentary reworking of beds is common. The greater development of Lake Superior type iron formation in early Proterozoic time has been considered by some to be related to increased atmospheric oxygen content, resulting from biological evolution.
ORE CONTROLS, GUIDES TO EXPLORATION	1. Distribution of iron formation is reasonably well known from aeromagnetic surveys. 2. Oxide facies is the most important, economically, of the iron formation facies. 3. Thick primary sections of iron-formation are desirable. 4. Repetition of favourable beds by folding or faulting may be an essential factor in generating widths that are mineable (30 to 150 m). 5. Metamorphism increases grain size, improves metallurgical recovery. 6. Metamorphic mineral assemblages reflect the mineralogy of primary sedimentary facies. 7. Basin analysis and sedimentation modelling indicate controls for facies development, and help define location and distribution of different iron formation facies.
AUTHOR	G.A. Gross

(see Plate 1, page 5)

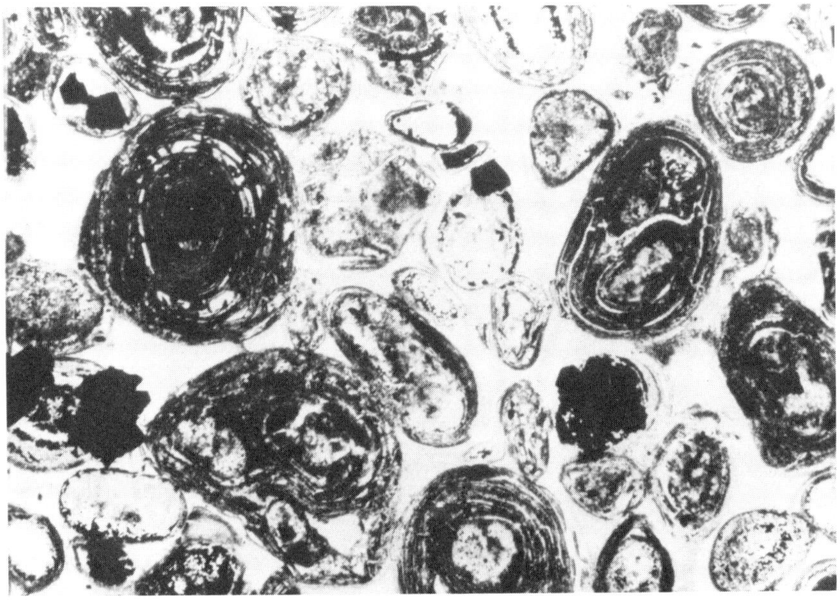

Figure 3. *2.2 Iron Formation (Lake Superior type). Lac Mistassini area, Quebec. Magnetite-hematite-quartz iron formation, oxide facies with oolitic texture. Field of view is 5 mm wide. Photo: G.A. Gross (GSC 203886-K).*

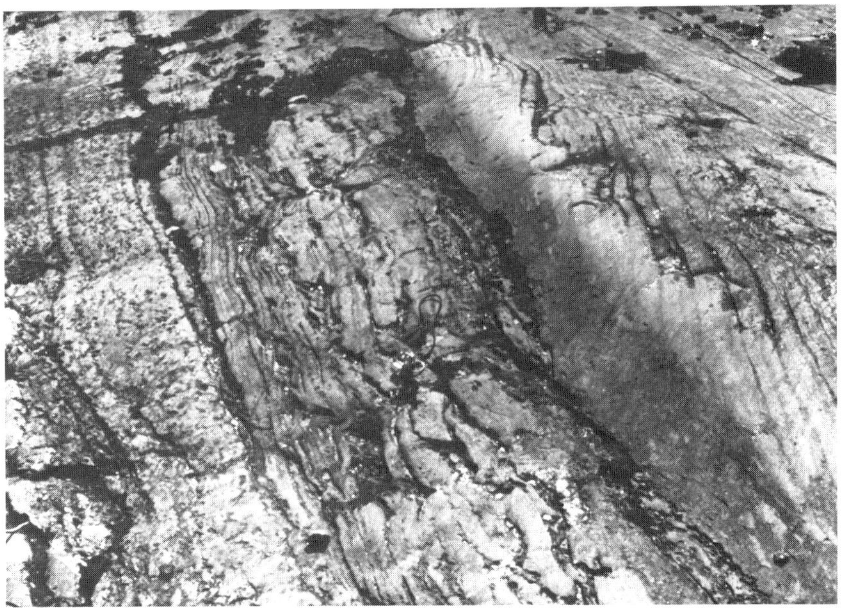

Figure 4. *2.2 Iron formation (Lake Superior type). Wabush Lake, Labrador. Metamorphosed specular hematite-quartz iron formation showing soft sediment deformation and lenticular bedding on left, stratigraphically below crossbedded unit. Magnet in centre is 3.5 cm. Photo: G.A. Gross (GSC 203886-U).*

Figure 5. *2.3 Iron Formation (Algoma type). Adams mine, Kirkland Lake, Ontario. Typical magnetite-chert iron formation, highly deformed. Pocket knife in centre is 10 cm. Photo: G.A. Gross (GSC CN-98).*

2.3 IRON FORMATION (ALGOMA TYPE)

COMMODITIES

Fe (Mn)

EXAMPLES:
Canadian – *Foreign*

Helen Mine at Wawa, Sherman Mine at Temagami, Griffith Mine at Ear Falls, and Lake St. Joseph, Ont.; Woodstock, N.B. – *Krivoy Rog, U.S.S.R.*

IMPORTANCE

Canada: second most important (after Lake Superior type) as a source of iron. Potential source of manganese (Woodstock).

TYPICAL GRADE, TONNAGE

Up to billions of tonnes, with grades ranging from 15 to 45% Fe, averaging 25% Fe. Manganese content is generally low in Precambrian deposits (generally less than 2%) but is more significant in some Paleozoic deposits (Mn=10 to 40%). Fe:Mn may range from 40:1 to 1:50.

GEOLOGICAL SETTING

Iron formation members occur with volcanic rocks, greywacke and shale near or distal from extrusive centres, along volcanic belts, deep fault systems, and rift zones; may be present at any stage in a volcanic succession. Most abundant in Archean greenstone belts. Some oxide, carbonate and sulphide facies have polymetallic sulphide facies associated with them.

HOST ROCKS OR MINERALIZED ROCKS

Oxide, silicate, carbonate and sulphide facies of banded iron-formation are commonly composed of thin, alternating layers or beds of silica (chert and quartz) and iron-rich minerals; and are interbedded with clastic sedimentary and volcanic strata.

ASSOCIATED ROCKS

Felsic, mafic and ultramafic volcanic rocks, greywacke, black shale, argillite, chert, interlayered pyroclastic and other volcaniclastic beds; metamorphic equivalents.

FORM OF DEPOSIT, DISTRIBUTION OF ORE MINERALS

Ores are sedimentary sequences commonly up to 100 m thick, and several kilometres in strike length. In most cases, isoclinal folding or thrust faulting have produced thickened sequences of iron formation, thus greatly enhancing economic mining feasibility. Ore mineral distribution closely reflects primary sedimentary facies.

MINERALS: Principal ore minerals
– *Associated minerals*

Magnetite, hematite; siderite, manganoan siderite, pyrite, and pyrrhotite are mined in a few deposits.
– Chert, quartz, Fe-silicates and -carbonates, chlorite, amphiboles, biotite, feldspar, garnet, chalcopyrite

AGE, HOST ROCKS

Precambrian to Recent, but predominantly Archean.

AGE, ORE

Syngenetic, same age as host rocks.

GENETIC MODEL

Chemical and colloidal precipitation of iron and silica in euxinic and oxidizing environments; iron and silica derived from volcanic effusive and hydrothermal sources along volcanic belts and deep fault or rift systems. Formation and distribution evidently controlled by tectonic rather than by biogenic or atmospheric factors.

ORE CONTROLS, GUIDES TO EXPLORATION

1. Distribution of iron formation is reasonably well known from aeromagnetic surveys.
2. Oxide facies is the most favourable, economically, of the iron formation facies.
3. Thick primary beds (30 to 100 m) of iron formation are desirable.
4. Repetition of favourable beds by folding or faulting is economically favourable.
5. Metamorphism increases grain size, improves metallurgical recovery.
6. Metamorphic mineral assemblages reflect the mineralogy of primary sedimentary facies.
7. Basin analysis and tectonic and sedimentation modelling indicate controls for facies development, and help define location and distribution of different iron formation facies.

AUTHOR

G.A. Gross

3. ENRICHED IRON FORMATION

COMMODITIES
Fe (Mn)

EXAMPLES:
Canadian – *Foreign*
French, Ruth and Timmins Mines of the Knob Lake-Schefferville area, Que. and Lab.; Atikokan, Ont. – *Mesabi and Cayuna Ranges, Minnesota; Marquette Range, Michigan*

IMPORTANCE
Canada: these deposits were the initial basis of development in the Knob Lake-Schefferville area, but are now of diminishing importance.
World: Major source of iron in Australia, Brazil, U.S.S.R. and Africa and previously in U.S.A.

TYPICAL GRADE, TONNAGE
1 to 50 million tonnes in Canada, up to billions of tonnes in Brazil and Australia. Fe content ranges from 45 to 69% Fe, averaging 50 to 60% Fe. Mn content generally low but variable.

GEOLOGICAL SETTING
Deep chemical weathering of both Lake Superior and Algoma type iron formation protore.

HOST ROCKS OR MINERALIZED ROCKS
Oxidized and chemically enriched residual zones of porous, friable iron-formation and earthy hydrated iron oxide, developed on iron formation protore.

ASSOCIATED ROCKS
Unaltered iron formation; shale, siltstone, quartzite, dolomite, and other clastic sedimentary and pyroclastic rocks, commonly oxidized and leached.

FORM OF DEPOSIT, DISTRIBUTION OF ORE MINERALS
Irregular lenses to tabular masses within iron formation protore. Ore mineral assemblages reflect the protore facies from which they were derived, i.e. blue, yellow, and red ore types were derived from oxide, silicate-carbonate, and ferruginous clastic facies, respectively (e.g. Knob Lake); in some deposits, brown goethite ores were derived from mixed carbonate-sulphide facies (e.g. Atikokan). Enriched ore tends to grade into an oxidized and partially leached protore.

MINERALS: Principal ore minerals – *Associated minerals*
Hematite, goethite, magnetite, martite, siderite, psilomelane, pyrolusite, hollandite.
– *Quartz, chert, iron silicate, dolomite, various rock-forming silicates, especially clay minerals*

AGE, HOST ROCKS
Age of protore: principally Precambrian.

AGE, ORE
Poorly documented; various periods of enrichment, Precambrian to Recent.

GENETIC MODEL
Residual iron oxide masses formed by secondary enrichment processes involving oxidation of the iron and leaching of the siliceous minerals through the action of deeply circulating groundwater in fractured and more permeable parts of iron formations.

ORE CONTROLS, GUIDES TO EXPLORATION
1. Upper, deeply chemically weathered surface of exhumed Precambrian iron formation.
2. Chemical composition, mineralogy and texture of primary facies of iron formation.
3. Grade of metamorphism of protore: grain size and texture affect permeability and hence oxidation and weathering.

AUTHOR
G.A. Gross

(see Plate 3, page 6)

4. STRATIFORM PHOSPHATE (PHOSPHORITE)

4.a Miogeosynclinal
4.b Platformal

COMMODITIES

P (U, F, V, artificial gypsum)

EXAMPLES:
Canadian – *Foreign*

(4.a) Fernie Synclinorium (Jurassic), B.C. and Alta.; Rapid Creek-Big Fish River area (Cretaceous), Yukon – *Phosphoria Formation (Permian), Idaho, Montana and Wyoming*
(4.b) Athabasca Basin (Proterozoic), Sask. – *Florida (Tertiary); Georgina Basin (Cambrian), Australia; Northwest Africa (Cretaceous-Eocene)*

IMPORTANCE

Canada: no past or current production. However, occurrences are known (see above), and other areas may have geological potential.
World: Stratiform phosphate, mainly subtype 4.b, accounts for more than 80% of world phosphate production.

TYPICAL GRADE, TONNAGE

Foreign: Average grades are typically 31 to 36% P_2O_5. Grades as low as 24% can be mined, and beneficiated to greater than 30%. Tonnages of deposits in the U.S.A. are of the order of hundreds of millions or billions of tonnes at grades greater than 24%. Tens to hundreds of ppm U in some deposits.

GEOLOGICAL SETTING

(4.a) Miogeosynclinal: in the miogeosynclinal wedge, especially in the flexure zone between the shallower shelf and the deeper basin, or where the sedimentary section consists of reworked sediments, and is therefore reduced in thickness.
(4.b) Platformal: usually in cratonic basins and stable shelves bordering cratons.
An oceanic plateau setting is also known, but is quantitatively minor.

HOST ROCKS OR MINERALIZED ROCKS

"Phosphorite" is a sedimentary rock that may be sandy, carbonate-rich or shaly and contains an abundance of marine phosphate (apatite) as pellets or nodules, or in the matrix.

ASSOCIATED ROCKS

Other sedimentary rocks characteristic of the depositional environment:
(4.a) chert, carbonaceous shale, carbonate.
(4.b) carbonate, sandstone, siltstone.

FORM OF DEPOSIT, DISTRIBUTION OF ORE MINERALS

Tabular, bedded; individual beds may be from less than one to several metres thick and extend over hundreds of thousands of square kilometres. The mineable parts of these beds are those for which surface mining methods are feasible.

MINERALS: Principal ore minerals
– Associated minerals

Apatite group minerals.
– Quartz, clays, carbonates

AGE, HOST ROCKS

Canada: Proterozoic to Cretaceous.
World: Proterozoic to Holocene.

AGE, ORE

Syngenetic, diagenetic; essentially the same age as the host rocks.

GENETIC MODEL

Upwelling of phosphate-rich oceanic waters onto shelves. Biogenic deposition in marine shelf and cratonic basin environments, accompanied by a warm, arid climate, and a slow rate of sedimentation of terrigenous material; diagenesis and later enrichment; upgrading by sedimentary reworking is important at many localities.

ORE CONTROLS, GUIDES TO EXPLORATION

(4.a), (4.b) the lithological associations of marine shelf and cratonic basin environments, e.g. chert, black shale; stratigraphic sections that are reduced in thickness due to reworking of sediments; low paleolatitudes.
(4.b) paleotopographic structural associations, e.g. structures representing former shallow topographic domes and basins.
The common association of uranium with stratiform phosphate makes radiation-detecting methods useful in exploration.

AUTHOR

R.L. Christie

5. PLACER URANIUM, GOLD

5.1.a Uranium
5.1.b Gold

COMMODITIES

(5.1.a) U (Th, rare-earth elements)
(5.1.b) Au, U (diamonds, Pt group metals)

EXAMPLES: Canadian – *Foreign*

(5.1.a) Elliot Lake, Ont.; Padlei, N.W.T.; Sakami Lake, Que. – *Medicine Bow Mountains, Wyoming; Black Hills, South Dakota*
(5.1.b) – *Witwatersrand, South Africa; Jacobina, Brazil*

IMPORTANCE

(5.1.a) Canada: constitutes one half of 1978 domestic uranium reserves.
World: subtypes (5.1.a) and (5.1.b) together account for one-third of 1978 non-Communist world uranium reserves.
(5.1.b) Canada: no reserves.
World: accounts for about 60% of total world reserves and annual production of gold, almost entirely from Witwatersrand, South Africa.

TYPICAL GRADE, TONNAGE

Tonnage of individual ore bodies generally falls in the range 10 to 400 million tonnes.
(5.1.a) Elliot Lake: grade average 900 ppm U (rich ore averages about 1200 ppm U, 250 ppm Th, 0.2 ppm Au, 2 ppm Ag; "downstream" marginal areas and upper ore beds average about 500 ppm U, 1000 ppm Th). Agnew Lake (near Elliot Lake): grade averages 700 ppm U and 2500 ppm Th.
(5.1.b) Witwatersrand: grade averages 10 ppm Au, 280 ppm U, 30 ppm Ag.

GEOLOGICAL SETTING

Occur in quartzose arenites in the oldest supracratonic successions deposited in intracratonic rift zones, intracratonic basins, or possibly proximal parts of marginal cratonic basins that rest on Archean rocks. The host successions lack redbeds, and predate any successions that contain undisputed redbeds. Important iron formations are found in overlying or nearby younger marine successions in South Africa, Australia, Wyoming, and the eastern Canadian Shield.

HOST ROCKS OR MINERALIZED ROCKS

Quartz pebble conglomerate and pebbly grit in coarse fluviatile sandstone.
(5.1.a) Uranium occurs mainly in quartz pebble conglomerate beds and with heavy minerals in pyritic layers in arenite.
(5.1.b) Gold and uranium occur in quartz pebble conglomerate beds, and in thin hydrocarbon seams and pyritic heavy mineral layers in quartzose arenite.

ASSOCIATED ROCKS

Argillaceous and quartzose or feldspathic sandstone. Also commonly present are polymict paraconglomerate, siltstone and shale, orthoconglomerate with clasts of quartz and basalt, gabbroic sheet intrusions and basalt. (Other associated rocks: at Jacobina, Brazil, serpentinite. At Sakami Lake, Quebec, serpentinite, ultramafic flows, turbiditic sediments, mafic and felsic volcanics, and iron formation. At Black Hills, South Dakota, hematite-bearing iron formation).

FORM OF DEPOSIT, DISTRIBUTION OF ORE MINERALS

Extensive stratiform single or multiple lenses, ribbon- or fan-shaped in plan.
(5.1.a) Uraniferous pyritic quartz pebble conglomerate; uraniferous heavy mineral layers in arenite near conglomerate beds.
(5.1.b) Auriferous and uraniferous quartz pebble conglomerate; U- and Au-bearing hydrocarbon seams in nonconglomeratic quartzite.

MINERALS: Principal ore minerals – *Associated minerals*

Uraninite, brannerite, uranothorite, uranoan monazite, native gold.
– Quartz as pebbles and smaller clasts, pyrite grains, interstitial sericite, K-feldspar grains, rutile-anatase, monazite, zircon, other heavy minerals including platinum group minerals and diamonds

AGE, HOST ROCKS

Early Aphebian and Archean. None known to be younger than 2.2 Ga. South African host rocks range from 2.9 to 2.2 Ga. Huronian host rocks in Canada are bracketed between events at 2.7 and 2.25 Ga. Australian host rocks are about 2.3 to 2.4 Ga. Sakami Lake host rocks are older than 2.4 Ga. Others (Brazil, Padlei, Wyoming, Black Hills) are bracketed between events at 2.5 and 1.8 Ga.

AGE, ORE

Same as host rocks, but significant diagenetic modifications may have occurred, including formation of brannerite in titania grains, and perhaps precipitation of gold.

GENETIC MODEL

Placer concentrations, formed prior to the development of an oxygenic atmosphere, of heavy ore minerals or their pre-diagenetic precursors along with detrital pyrite, iron and titanium oxides, monazite, zircon, and other heavy minerals.

ORE CONTROLS, GUIDES TO EXPLORATION

Paleotopographic features that produced abrupt changes in gradient and competence of streams that carried detritus derived from weathered granitoid terranes (e.g. valleyheads, fault scarps, lava field margins, bajadas, and stream confluences).

Exploration guides: Examine any coarse grained, quartz-rich arenite that may be older than 2.25 Ga or of fluvial origin and contains pyrite but no hematite. If there are concentrations, however slight, of radioactivity in coarser layers, seek higher energy environments (pebbly beds) within the succession. Explore such beds up the paleocurrent direction seeking coarsening of pebbles, thickening of pebble beds and increase in U:Th ratios.

AUTHOR

S.M. Roscoe

(see Plates 4, 5, page 7)

5.2 PLACER GOLD

COMMODITIES
Au (Ag)

EXAMPLES: Canadian – *Foreign*
Klondike District, Yukon; Cariboo District, B.C.; Chaudière, Que. – *Sierra Nevada, California; Victoria, Australia; Lena, Aldan and Amur Rivers, U.S.S.R.; Choco, Colombia*

IMPORTANCE
Canada: Accounted for 8% of total recorded Canadian gold production to 1955. Declined from 2% of annual Canadian gold production in 1954 to 0.2-0.3% in late 1960s and early 1970s, rising again to over 3% in 1980.
World: Estimated to account for one fourth to one third of total past production.

TYPICAL GRADE, TONNAGE
Deposits range from high grade "bonanzas" to low grade, non-economic disseminations. In the Klondike and Cariboo districts, many tens of kilometres of gold-bearing valley gravels were exploited, perhaps the most exceptional being Eldorado Creek (Klondike) reportedly averaging 600 g Au/metre for 6 km.

GEOLOGICAL SETTING
Deeply weathered unglaciated terrane (commonly part of a stable craton, containing suitable auriferous host rocks) having moderately incised stream valleys containing various types of alluvial deposits.

HOST ROCKS OR MINERALIZED ROCKS
Gulch, creek, river, flood plain, deltaic, beach and near-shore sand and gravel beds, or elevated equivalents; may be consolidated to form paleoplacers.

ASSOCIATED ROCKS
The bedrock, which includes the source rocks for Au, is highly varied. Metamorphic rocks are especially common, and include a wide variety of metasedimentary schists and gneisses, as well as volcanic, ultramafic and granitic rocks, Precambrian to Tertiary in age.

The probable source rocks for most placer gold include gold-bearing quartz veins, felsic intrusions, base metal sulphide deposits, paleoplacer deposits, and other gold-bearing deposits.

FORM OF DEPOSIT, DISTRIBUTION OF ORE MINERALS
Ranges from narrow pay streaks on or near bedrock in narrow valleys, to zones or layers of sparsely disseminated very fine grained gold in flood plain, deltaic, beach or near shore sands and gravels. Grain size decreases with distance from source.

MINERALS: Principal ore minerals – *Associated minerals*
Native gold, electrum, platinum.
– *Quartz, magnetite, ilmenite, hematite, garnet, zircon, pyrite and various other locally derived heavy minerals*

AGE, HOST ROCKS
Chiefly Tertiary, Quaternary. Some older paleoplacers.

AGE, ORE
Same as host sediments. Derived from source rocks of Precambrian to Tertiary age.

GENETIC MODEL
Deep secular weathering of suitable source rocks results in preconcentration of gold on or near bedrock surface by gravity, creep, frost action and solifluction. Further concentration accompanies uplift and attendant incision of stream valleys and formation of alluvial deposits. Gravity and stream action concentrate gold into paystreaks in the alluvium, especially on naturally riffled bedrock surface, but also on impervious layers within alluvium. Additional uplift and downcutting may leave bench placers (stream terraces) or result in redistribution into river, floodplain, deltaic, and beach placers.

Some gold in placer deposits may be contributed by chemical migration and accretion processes.

ORE CONTROLS, GUIDES TO EXPLORATION
1. Gold bearing source rocks.
2. Evidence of deep weathering of source rocks.
3. Absence of, or protection from, extensive glaciation (i.e., preservation of weathered profile before 4.).
4. Superposed, moderately incised drainage.
5. Favoured sites in resultant alluvial deposits are concentrations on "riffled" bedrock surface or impervious layers, or in zones within flood plain, deltaic, beach or near-shore sand and gravel beds.
6. Coarser gold is commonly found in upper reaches of streams, and on steeper gradients; the richest and coarsest gold concentrations typically occur in comparatively coarse sediment.

AUTHOR
C.R. McLeod

Figure 6. *5.2 Placer Gold. Klondike area, Yukon. Basal part of high level White Channel bench gravels. Early placer miners (c.1902) followed lower bedrock surface with timbered drift (about 1.2 m wide). Bench mined hydraulically and with bulldozer in 1960. Photo: C.R. McLeod (GSC 202282-V).*

Figure 7. *6.1 Mississippi Valley Lead-Zinc. Pine Point, Northwest Territories. Dolomite collapse breccia cemented by sphalerite, galena, pyrite, marcasite, and dolomite. Some sulphide cement shows texture similar to that in Plate 2. Photo: D.F. Sangster (GSC 202282-U).*

6.1 MISSISSIPPI VALLEY LEAD-ZINC

COMMODITIES

Pb, Zn (Ag, Cd)

EXAMPLES:
Canadian – *Foreign*

Pine Point and Polaris, N.W.T.; Newfoundland Zinc, Nfld.; – *Viburnum Trend and Old Lead Belt Districts, Missouri; East Tennessee District, Tennessee; Silesian District, Poland*

IMPORTANCE

Canada: about 30% of lead-zinc production.
World: major source of lead and zinc in U.S.A., Poland and Austria.

TYPICAL GRADE, TONNAGE

Data for individual deposits are difficult to obtain because of lack of production records and the fact that, in many districts, deposits tend to be interconnected. Best estimate for most individual deposits: 5 to 10% combined Pb-Zn, 1 to 10 million tonnes.

GEOLOGICAL SETTING

In platform carbonate successions. Commonly, but not always, located between a zone of tectonic instability characterized by vertical movement (commonly called a "hinge line" and marked by rapid lithological facies changes such as at a reef front, or edge of a sedimentary basin), and the tectonically stable platform.

HOST ROCKS OR MINERALIZED ROCKS

Carbonate rocks, generally highly brecciated dolomite.

ASSOCIATED ROCKS

Most commonly limestone; less commonly shale, sandstone and evaporites.

FORM OF DEPOSIT, DISTRIBUTION OF ORE MINERALS

Form: highly irregular in shape, usually discordant on deposit-scale but stratabound on a district scale.
Distribution of ore minerals: mostly as open-space filling in highly brecciated dolomite in which sphalerite, especially, shows colloform texture. Also commonly disseminated with secondary carbonate gangue; occasionally massive, coarsely crystalline aggregates.

MINERALS: Principal ore minerals
– *Associated minerals*

Sphalerite, galena.
– *Pyrite, marcasite, dolomite, calcite, lesser amounts of quartz, barite, fluorite, chalcopyrite*

AGE, HOST ROCKS

Canada: Helikian to Carboniferous; most abundant in early to mid-Paleozoic.
Foreign: mainly Cambrian to Triassic.

AGE, ORE

Not known with any certainty.

GENETIC MODEL

Although "...no general consensus has been reached...geologists do not know the physiochemical reasons why Mississippi Valley deposits are where they are..." (Ohle, 1980, p. 163), fluid inclusion studies suggest ores were precipitated from low temperature (commonly 80°C - 150°C) brines. A commonly cited model is based on the Beales and Jackson (1966) interpretation whereby the brines originated from shale basins adjacent to the platform carbonates, and the ore minerals precipitated in some cases during early diagenesis, and in other cases long after lithification of host rocks.

ORE CONTROLS, GUIDES TO EXPLORATION

No consensus on genetic models hence no consensus on ore controls, or guides. However, one or more of the following features are commonly associated with these deposits:

1. Secondary breccia in dolomite, cemented by white sparry dolomite.
2. Unconformities within carbonate sequence (ore horizon will be below unconformities).
3. Reefs.
4. Carbonate-shale and limestone-dolomite facies changes.
5. Basement high(s).
6. Open spaces of any type within carbonate sequences, especially those formed by karstification as evidenced by brecciation, thinning of carbonate strata, local increase in concentration of insoluble residue material.

AUTHOR

D.F. Sangster

(see Plate 2, page 5)

6.2 SANDSTONE LEAD

COMMODITIES	Pb (Zn, Ag, Cu, As, Ni, Co)
EXAMPLES: Canadian – ***Foreign***	Yava, N.S.; George Lake, Sask. – *Laisvall, Sweden; Maubach-Mechernich, West Germany; Largentière, France; Zeida, Morocco*
IMPORTANCE	Canada: only of local importance. World: major importance in Sweden, somewhat less in France. Major production in Germany during World War II.
TYPICAL GRADE, TONNAGE	2 to 5% Pb, 0.2 to 0.8% Zn, 1 to 20 g Ag/tonne mostly less than 10 million tonnes, some up to 80 million tonnes. Unlike other deposits in this subtype, George Lake is zinc-dominant rather than lead-dominant.
GEOLOGICAL SETTING	Transgressive basal quartzitic or quartzo-feldspathic sandstones resting on deeply weathered sialic basement (in three districts, basement has been shown to be geochemically enriched in lead, i.e. averaging greater than 40 ppm). Generally low paleolatitude conditions prevailed.
HOST ROCKS OR MINERALIZED ROCKS	Clean, quartzitic or quartzo-feldspathic sandstones; conglomerates.
ASSOCIATED ROCKS	Sialic basement (e.g., granitic rocks, rhyolite), evaporites, siltstone, shale, minor carbonate (dolomite).
FORM OF DEPOSIT, DISTRIBUTION OF ORE MINERALS	Usually lensoid, broadly conformable to bedding in sandstone. Internally, sulphides are disseminated in homogeneous sandstone or concentrated in high grade zones along bedding. Concretionary and poikiloblastic sulphide concentrations are abundant in some deposits. Remobilization into faults is a common feature.
MINERALS: Principal ore minerals – ***Associated minerals***	Galena, sphalerite. – *Secondary silica (commonly chalcedony), carbonate, pyrite, organic material, chalcopyrite, minor sulphates, sulphosalts and other sulphides*
AGE, HOST ROCKS	Various ages from Aphebian to Cretaceous.
AGE, ORE	Same as host rocks or slightly younger.
GENETIC MODEL	Groundwater transport of metals leached from lead-rich basement, through porosity channels in sandstone; precipitation of metals by biogenically-produced sulphide. Recently, a genetic model involving compaction of brine-bearing basins by over-riding nappes has been proposed for deposits in Sweden.
ORE CONTROLS, GUIDES TO EXPLORATION	1. Sialic basement with high lead content (>30 ppm). 2. Basal quartz sandstone of a transgressive sequence, overlying basement. 3. Channels in sandstone as evidenced by thickening, lateral conglomerate-to-sandstone facies changes, etc. 4. Permeable zones in sandstone (i.e., "cleanest" sandstone, minimum of intergranular clayey material).
AUTHOR	D.F. Sangster

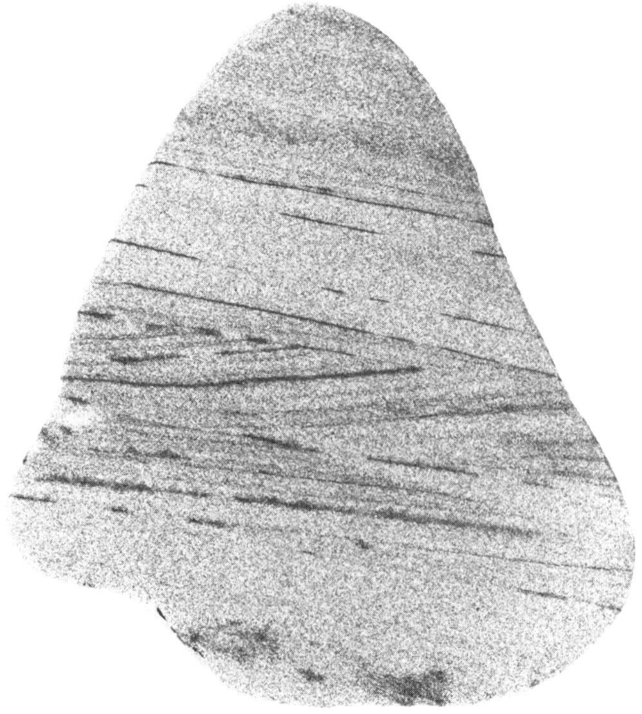

Figure 8. *6.2 Sandstone Lead. Yava deposit, Nova Scotia. Galena and carbonaceous material (mainly wood trash) oriented parallel to bedding in crossbedded terrigenous sandstone. Wood trash is detrital; galena occurs as cement between quartz grains. Sample is approximately 20 cm high. Sample: R.V. Kirkham (GSC 203641-S).*

6.3 SEDIMENTARY COPPER

6.3.a Paralic marine (Kupferschiefer-type)
6.3.b Continental (Red bed-type)

COMMODITIES	Cu (Ag, Co)
EXAMPLES: **Canadian – *Foreign***	**(6.3.a)** Redstone, N.W.T. – *Kupferschiefer, Poland-Germany; Zambian and Zairean Copperbelts; Udokan, U.S.S.R.; White Pine, Michigan; Spar Lake, Montana; Creta, Oklahoma.* **(6.3.b)** Dorchester, N.B. – *Dzhezkazgan, U.S.S.R.: Nacimiento, New Mexico.*
IMPORTANCE	Canada: No economic deposits in Canada but large deposits in United States near Canadian border. World: 15% to 20% of world copper production and reserves; primarily from a few large districts such as Zambian and Zairean Copperbelts; Lubin, Poland; and Dzhezkazgan, U.S.S.R.
TYPICAL GRADE, TONNAGE	**(6.3.a)** Highly variable. 1.0 to 5.0% Cu and 1 to 30 g Ag/tonne; 5 to 500 million tonnes. Cobalt is an important byproduct in Zambian and Zairean Copperbelts. **(6.3.b)** 1 to 2% Cu and 1 to 30 g Ag/tonne, 1 to 10 million tonnes.
GEOLOGICAL SETTING	Continental or shallow marine sedimentary rocks deposited in low latitude, arid and semi-arid environments. Evaporites occur in the section. **(6.3.a)** Anoxic marine rocks overlie or are interlayered with redbeds. **(6.3.b)** Anoxic fluvial and lacustrine rocks overlie or are interlayered with redbeds.
HOST ROCKS OR MINERALIZED ROCKS	**(6.3.a)** Carbonaceous claystone, siltstone, sandstone, marl, limestone and dolomite. **(6.3.b)** Carbonaceous sandstone, conglomerate, claystone and siltstone.
ASSOCIATED ROCKS	Redbeds, evaporites.
FORM OF DEPOSIT, DISTRIBUTION OF ORE MINERALS	Concordant or peneconcordant zones of disseminated sulphides, mainly tabular or blanket-shaped, but also channel-like or linear. Typical lateral extent of mineralized beds is of the order of kilometres; typical thickness, 0.5 to 30 m. Sulphides are commonly zoned both vertically and laterally, showing part or all of the following sequence (upward and outward from the base of the orebody): native copper, chalcocite, bornite, chalcopyrite, galena, sphalerite, pyrite.
MINERALS: Principal ore minerals ***– Associated minerals***	Chalcopyrite, bornite, chalcocite, native copper, carrollite. *– Pyrite, other sulphides, ordinary rock forming minerals of sedimentary rocks such as quartz, feldspar, carbonates, clays*
AGE, HOST ROCKS	Early Proterozoic (about 2.25 Ga) to Tertiary. Only after the formation of undisputed redbeds.
AGE, ORE	The same as, or slightly younger than, host rocks.
GENETIC MODEL	Diagenetic subsurface brines (probably derived from evaporites) extracted copper from available basement rocks or sediments, transported it through oxidized beds, and precipitated it by reduction in anoxic sediments. Early diagenetic pyrite was a common reductant.
ORE CONTROLS, GUIDES TO EXPLORATION	1. Low latitude, arid, continental and shallow marine sedimentary sequences. 2. Sources of copper such as copper-bearing basement and/or sediments. 3. Extensive redbed or other oxidized aquifer system and adjacent pyritic, carbonaceous host rocks. The typical sites of ore deposition differ in the two subtypes: **(6.3.a)** the base of a major marine transgressive unit overlying redbeds; **(6.3.b)** the permeable lower parts of fining-upwards fluvial cycles. 4. Large-scale zoning of sulphides (indicating that the mineralizing systems were large-scale phenomena).
AUTHOR	R.V. Kirkham

Figure 9. *6.3.a Sedimentary Copper (Kupferschiefer type). Redstone deposit, Northwest Territories. Typical pale, fine grained silty carbonate grainstone with delicate climbing ripples and erractically disseminated chalcopyrite. The 12 mm-long lens (lower centre) is a calcite nodule with chalcopyrite along the top. Sample is 12 cm long. Sample: R.V. Kirkham (GSC 203641-W).*

6.4 SANDSTONE URANIUM

COMMODITIES

U (V, Mo, Se)

EXAMPLES: Canadian – *Foreign*

Blizzard deposit, Kelowna, B.C.; Mountain Lake, N.W.T. – *Lucky Mac Mine, Wyoming; Jackpile Mine, New Mexico*

IMPORTANCE

Canada: mineable deposits in B.C. account for less than 2% of Canada's "Reasonably Assured Resources"* of uranium.
World: accounts for 40% of world's "Reasonably Assured Resources"* and 54% of "Estimated Additional Resources"* of uranium. Contained V, Mo and Se are of minor or negligible importance.

TYPICAL GRADE, TONNAGE

Canada: up to about 4000 tonnes of contained U in ore bodies grading 0.1 to 0.2% U.
Foreign: the common range in mined deposits is 1000 to 10 000 tonnes of contained U in ores grading 0.1 to 0.2% U. Many deposits contain less than 1000 tonnes U, but a few contain more than 30 000 tonnes U. Some uranium deposits in western U.S.A. contain up to 1.5% V_2O_5, and some up to 0.2% Mo (Mo in many cases is deleterious).

GEOLOGICAL SETTING

Mainly in closed sedimentary successor basins adjacent to dominantly continental hinterlands containing granitic rocks and/or felsic volcanic rocks. Generally the basins contain post-Devonian clastic sedimentary sequences with tuffs.

HOST ROCKS OR MINERALIZED ROCKS

Clastic continental sedimentary rocks, predominantly semi-consolidated sandstone and conglomerate containing carbonaceous matter.

ASSOCIATED ROCKS

Tuff commonly occurs in the sedimentary sequence. In the Blizzard and nearby deposits, the host rocks rest unconformably on igneous and/or metamorphic rocks and are overlain by plateau basalts.

FORM OF DEPOSIT, DISTRIBUTION OF ORE MINERALS

Most common forms of deposits are
(1) roll ("c"-shape in cross-section),
(2) blanket, and
(3) stack (ranges from interconnected, vertically stacked blankets, to near vertical "pipes").

The Blizzard deposit is of blanket or ribbon form following paleoriver channels. The ore minerals occur in the matrix and pores of the hosting sandstone.

MINERALS: Principal ore minerals – *Associated minerals*

Reduced zone
Pitchblende, coffinite.
– *Pyrite, marcasite; shiny, firm carbonaceous matter*

Oxidized zone
Autunite, carnotite, tyuyamunite.
– *Clay minerals secondary after feldspar, limonite, hematite; dull, soft carbonaceous matter*
In the Blizzard and nearby deposits, saléeite and autunite predominate, and ningyoite is known.

AGE, HOST ROCKS

Mainly post-Devonian, most commonly Mesozoic and Tertiary; uncommon examples in Proterozoic, e.g. Mountain Lake, N.W.T.

AGE, ORE

Younger than the host rocks. Mainly Mesozoic and Cenozoic.

GENETIC MODEL

Under relatively arid continental conditions, uranium was derived from felsic igneous rocks, both in the sediment source area and within the host sequence; transported in oxidized form in groundwaters through permeable clastic rocks; and deposited on encountering reducing conditions at a "redox front". Reductants were usually carbonaceous matter, sulphides or methane; phosphates (e.g. Blizzard) and vanadates also fix uranium. Bacterial activity likely played an important role in first leaching (low pH, high Eh), and then fixing (low Eh) uranium at "redox" fronts. Commonly formed during diagenesis of host sequence.

These deposits are considered by many to be genetically similar to unconformity type deposits (Type 21).

ORE CONTROLS, GUIDES TO EXPLORATION

1) The interface zone within loosely consolidated continental conglomerate and sandstone beds between the oxidized zone (characterized by hematite, limonite, clay minerals and dull, soft carbonaceous material) and the reduced zone (characterized by pyrite and other sulphides, feldspar, and shiny, firm carbonaceous matter).
2) The mineralized beds are bounded above and below, and in some cases laterally, by impermeable strata.
3) In the case of "stack"-type deposits, faults are important, having provided channelways for introduced reductants or for uranium-bearing waters.
4) Preservation of such deposits is enhanced if structural deformation is not severe (i.e., if dips are shallow), and if they are capped by impermeable strata (e.g., shales, massive volcanic strata).

AUTHORS

V. Ruzicka, R.T. Bell

---

* "Reasonably Assured Resources" = measured plus indicated reserves
"Estimated Additional Resources" = inferred reserves plus speculative resources
See Report EP 81-3, Department of Energy, Mines and Resources, 1981

7. CHEMICAL—SEDIMENT—HOSTED GOLD

7.a Carbonate-oxide iron formation

7.b Arsenical sulphide-silicate iron formation

7.c Stratiform pyrite

7.d Chert-sulphide

COMMODITIES

Au (Ag, Cu)

EXAMPLES: Canadian – *Foreign*

(7.a) Geraldton and Pickle Crow districts, Ont. – *Vubachikwe, Zimbabwe*
(7.b) Lupin Mine, Contwoyto Lake, N.W.T. – *Homestake, South Dakota.*
(7.c) Doyon-Silver Stack, Dumagami, Eagle, and Montauban, Que. – *Morro Velho, Brazil; Sons of Gwalia, Leonora District, Australia.*
(7.d) Detour Lake, Ontario

IMPORTANCE

Canada: Subtypes 7.a and 7.c are estimated to account for 12 to 13% of Canada's total cumulative gold production. Significant future producers (currently under development) are deposits of the arsenical sulphide-silicate iron formation (7.b) and chert-sulphide (7.d) subtypes.
World: The Homestake mine currently accounts for 25% of the gold produced in the United States.

TYPICAL GRADE, TONNAGE

The more significant deposits fall in the ranges 6 to 17 g Au/tonne, and 1 to 5 million tonnes. Detour Lake reportedly contains at least 25 million tonnes averaging 4.3 g Au/tonne together with 0.20% Cu and 5.1 g Ag/tonne.

GEOLOGICAL SETTING

Mainly in Archean greenstone belts near a major transition from volcanic to sedimentary rocks. The arsenical sulphide-silicate iron formation, subtype 7.b, occurs in dominantly sedimentary terrane, whereas the Detour Lake deposit of chert-sulphide subtype is in predominantly volcanic terrane.

HOST ROCKS OR MINERALIZED ROCKS

(7.a) Carbonate-oxide iron formation.
(7.b) Arsenical sulphide-silicate iron formation.
(7.c) Massive to disseminated pyritic units, felsic to mafic tuffaceous and agglomeratic rocks; in part, sulphide iron formation (Eagle Mine, Morro Velho). The Eagle Mine is in a purely sedimentary sequence (siltstone, shale, chert).
(7.d) Chert-sulphide strata.

ASSOCIATED ROCKS

(7.a) Mafic volcanic rocks, metasedimentary rocks including greywacke and arkose.
(7.b) Metasedimentary rocks including shale, siltstone, greywacke and amphibole-rich sedimentary units.
(7.c) Variety of Precambrian metavolcanic and metasedimentary rocks.
(7.d) Detour Lake; tholeiitic pillowed lavas, unit of intermediate tuffs, and an ultramafic sill (?).

FORM OF DEPOSIT, DISTRIBUTION OF ORE MINERALS

Stratiform. In part disseminated uniformly in the host units and in part in irregularly distributed to systematic, structurally controlled, minor quartz veins. Commonly thickened and structurally remobilized in fold hinges. The stratiform pyritic subtype contains disseminated to massive sulphide.

MINERALS: Principal ore minerals – *Associated minerals*

Native gold, Au tellurides in some deposits, chalcopyrite at Detour Lake.
– Pyrite, arsenopyrite, pyrrhotite, magnetite, hematite, quartz, carbonates; loellingite and cummingtonite and other amphiboles occur mainly in subtype 7.b

AGE, HOST ROCKS

Archean in general; Montauban, Quebec, Aphebian or Helikian.

AGE, ORE

Same as host rocks; possibly younger in some cases.

GENETIC MODEL

Not well documented. For subtype 7.b and for subtype 7.c in part (Eagle Mine, Morro Velho), the gold may be syngenetically precipitated in chemical sediments. Many of the chemical components have probably been introduced by hydrothermal exhalations. In other cases such as Geraldton and Detour Lake gold has been either locally redistributed or else introduced from a remote source and deposited in chemically favourable strata.

ORE CONTROLS, GUIDES TO EXPLORATION

Predominantly in Archean greenstone belts. Presence of iron formation, particularly carbonate-oxide, sulphide, or sulphide-silicate (± arsenides) facies, with or without chert.

AUTHORS

R.I. Thorpe, J.M. Franklin

8. CLASTIC—SEDIMENT—HOSTED GOLD

8.1 CARBONACEOUS SHALE/CARBONATE-HOSTED GOLD (CARLIN TYPE)

COMMODITIES	Au (As, Hg, Ag, Sb, Tl)
EXAMPLES: **Canadian – *Foreign***	– *Carlin, Getchell, Cortez, Gold Acres, Jerrett Canyon and others, all in Nevada*
IMPORTANCE	Canada: deposits that are unquestionably of this type are not presently known in Canada but they may be present. World: Major current and past producers in Nevada; deposits of this type will be of increasing importance in the U.S.A. because of recent discoveries in Nevada.
TYPICAL GRADE, TONNAGE	Canada: no data. Foreign: A number of individual deposits of about 5 million tonnes are known. The Carlin mine (Nevada) has produced about 10 million tonnes from 3 deposits and has reserves of about 6 million tonnes. Grades range from 1.0 to 10 g Au/tonne.
GEOLOGICAL SETTING	Host rocks to the Nevadan deposits were deposited in shelf-basin transitional environments, characterized by slow sedimentation. These rocks are presently allochthonous in thrust fault slices. Near many deposits they have been intruded by Tertiary felsic plutons in the Basin and Range structural-volcanic setting.
HOST ROCKS OR MINERALIZED ROCKS	Highly carbonaceous interbedded shale-carbonate sequences, in part bleached and altered, that include both bedded and replacement black cherts. Alteration consists of decarbonatization, silicification, argillization, pyritization, redistribution and local concentration of carbon, and near surface acid leaching and oxidation.
ASSOCIATED ROCKS	Granodiorite and other felsic plutonic rocks, altered felsic dykes, and skarns.
FORM OF DEPOSIT, DISTRIBUTION OF ORE MINERALS	In most cases, generally tabular, stratabound bodies that are irregular in detail. Some ore zones are transgressive (Getchell) and consist of gold concentrations associated with abundant carbonaceous material in fault zones. Sulphides are sparsely disseminated in most cases. Relatively massive realgar and orpiment are present at Getchell.
MINERALS: Principal ore minerals **– *Associated minerals***	Native gold. – *Pyrite, arsenopyrite, stibnite, realgar, orpiment, rare thallium minerals; the common rock forming minerals of shale-carbonate sequences, e.g., calcite, dolomite, clay minerals, chert, carbonaceous material, some alteration minerals*
AGE, HOST ROCKS	Cambrian to Devonian.
AGE, ORE	Some deposits are possibly Mesozoic; others possibly Tertiary.
GENETIC MODEL	Syngenetic or diagenetic enrichment of gold in highly carbonaceous sediments is presumed important, in part because the characteristic suite of elements seems indicative of a carbonaceous source. In most cases there is evidence of later hydrothermal redistribution and/or concentration related to epithermal activity associated with small Mesozoic and Tertiary felsic plutons. Possible further redistribution during near-surface oxidation and removal of carbonaceous material.
ORE CONTROLS, GUIDES TO EXPLORATION	1. Highly carbonaceous carbonate or shale-carbonate sequences may be the main ore control. 2. Presence of small felsic plutons that may have caused geothermal activity. 3. Other possibly favourable indicators are the following: black chert; realgar and orpiment; stibnite; and the barite or tungsten mineralization in the general area. 4. As and Hg (in some cases Tl, Sb) geochemical anomalies are considered highly favourable.
AUTHOR	R.I. Thorpe

8.2 TURBIDITE-HOSTED VEIN AND SHEAR ZONE GOLD

COMMODITIES	Au (Ag, W)
EXAMPLES: **Canadian – *Foreign***	Gold vein districts in the Meguma Group, N.S.; veins and mineralized shear zones in Yellowknife Supergroup, N.W.T. and in Superior Province – *Ballarat and Bendigo Districts, Australia*
IMPORTANCE	Canada: limited current production from Archean deposits. Forty-seven districts in Nova Scotia produced a total of about 35 million g (1 134 000 oz) of gold. World: Historically important in the Ballarat-Bendigo districts, Australia, accounting for about 12% of Australia's total past production.
TYPICAL GRADE, TONNAGE	Canada: the largest district in Nova Scotia (Goldenville) produced 550 000 tonnes at 12 g/tonne. Other districts generally produced less than 200 000 tonnes at similar or somewhat higher grade.
GEOLOGICAL SETTING	In turbidite sequences deposited on continental rises or in deep submarine troughs or basins.
HOST ROCKS OR MINERALIZED ROCKS	Greywacke-shale sequences.
ASSOCIATED ROCKS	Devonian granitic plutonic rocks in Nova Scotia.
FORM OF DEPOSIT, DISTRIBUTION OF ORE MINERALS	Veins concordant with bedding, cross veins, saddle reefs (concordant to discordant vein quartz in fold crests); all tend to be concentrated at or near the axes of major folds. Native gold disseminated in the quartz or associated with arsenopyrite and other minor sulphides. Veins show evidence of deformation.
MINERALS: Principal ore minerals **– *Associated minerals***	Native gold. – *Arsenopyrite, minor pyrite, galena, pyrrhotite, and other sulphides, quartz, chlorite*
AGE, HOST ROCKS	Early Ordovician in Nova Scotia, Silurian in Australia. Archean in the case of some small deposits near Yellowknife, N.W.T. and a few deposits in Superior Province.
AGE, ORE	Probably Devonian in Nova Scotia and Australia. Archean in N.W.T.
GENETIC MODEL	Deposition of gold related to dilatant structures. The source of the gold has not been definitely established, but a possibility is that it was scavenged from associated carbonaceous, sulphidic, arsenical shales. Devonian granite may have been important in mobilizing the gold in the Nova Scotia deposits. Only small diorite-gabbro intrusions that are probably unrelated to the deposits have been recognized in the Ballarat-Bendigo districts, Australia.
ORE CONTROLS, GUIDES TO EXPLORATION	Ore veins and masses are strongly controlled by dilatant zones formed during deformation (formation of upright open folds and associated faults and fractures in Nova Scotia). Associated carbonaceous, sulphidic and arsenical sedimentary rocks possibly important.
AUTHOR	R.I. Thorpe

Figure 10. *9.1 Volcanic-Associated Massive Sulphide. Millenbach mine, Noranda, Quebec. Chalcopyrite-pyrite stockwork ("stringer") ore (white in photograph) in chloritized quartz-feldspar porphyry (dark) about 3 m below massive ore of No. 14 lens. Field of view is about 1.5 m wide. Photo: John W. Lydon (GSC 203886-I).*

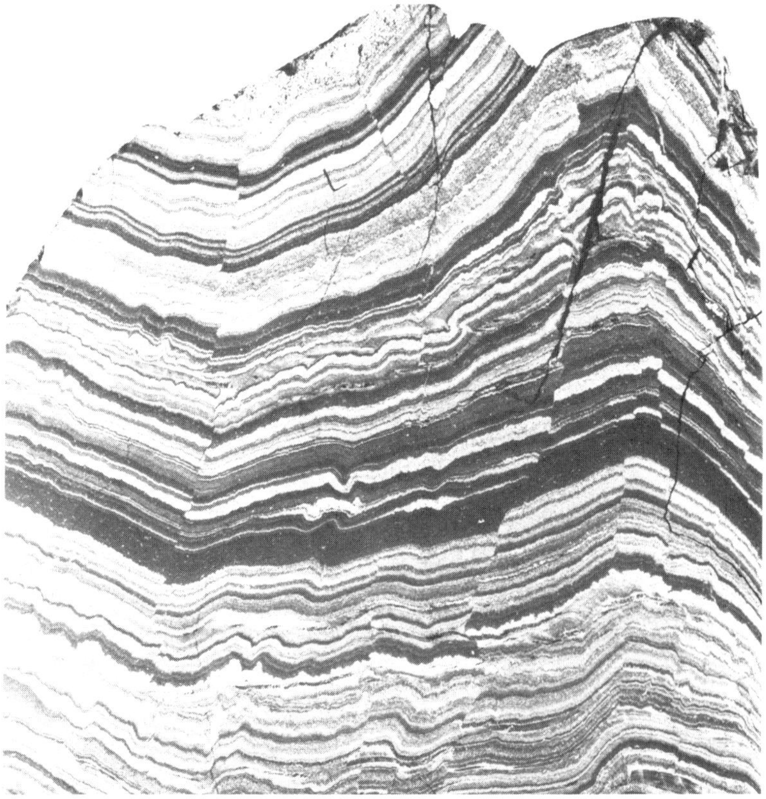

Figure 11. *9.2 Sediment-Hosted Sulphide. XY deposit, Howard's Pass, Yukon. Pyritic carbonaceous chert ore comprising thin beds of interlaminated beds of sphalerite-galena, framboidal pyrite, chert and cherty mudstone (sulphides are white, chert layers are dark). Width of field of view is 8 cm. Photo: I.R. Jonasson (GSC 204083-5).*

9. STRATIFORM SULPHIDE, BARITE

9.1 VOLCANIC-ASSOCIATED MASSIVE SULPHIDE

9.1.a Copper-Zinc
9.1.b Zinc-Lead-Copper

COMMODITIES

(9.1.a) Cu, Zn, Ag, Au (Pb, Cd, Sn)
(9.1.b) Zn, Pb, Cu, Ag (Au, barite, Cd, Sn, Sb, Bi)

EXAMPLES:
Canadian – *Foreign*

(9.1.a) Flin Flon Mine, Man-Sask.; Kidd Creek Mine, Ont; Millenbach and Horne Mines, Que.; York Harbour and Betts Cove Mines, Nfld. – *Norwegian Caledonide deposits, e.g. Lokken and Skorovas Mines; Cyprus deposits, e.g. Agrokipia*
(9.1.b) Brunswick No. 12, N.B.; Western Mines, B.C.; Buchans, Nfld. – *Kuroko deposits, e.g. Matsumine Mine, Japan; Spanish-Portugese Pyrite Belt deposits, e.g. Aljustrel Mine*

IMPORTANCE

(9.1.a) During 1977-1978, 28% of the copper, 41% of the zinc, 39% of the silver and 7.5% of the gold produced in Canada was from this type of deposit. Kidd Creek Mine alone accounted for 18% of Canadian silver production.
(9.1.b) During 1977-1978, 22% of the zinc, 27% of the lead, 2.5% of the copper and 19% of the silver produced in Canada was from this type of deposit.

TYPICAL GRADE, TONNAGE

(9.1.a) Average grade and tonnage of 52 deposits in the Abitibi Belt is 9.2 million tonnes containing 1.47% Cu, 3.43% Zn, 0.07% Pb, 31.7 g Ag/tonne and 0.81 g Au/tonne. Exclusive of the two "supergiants", Kidd Creek and Horne, the average tonnage of 50 deposits is 3.98 million tonnes but the average grade is approximately the same.
Average grade and tonnage of 38 deposits of the Norwegian Caledonides is 3.46 million tonnes containing 1.41% Cu, 1.53% Zn, 0.05% Pb (precious metal data not available).
(9.1.b) Average grade and tonnage of 29 deposits of Bathurst, N.B. camp is 8.70 million tonnes containing 0.56% Cu, 5.43% Zn, 2.17% Pb, 60.03 g Ag/tonne and 0.47 g Au/tonne. Exclusive of the "supergiant" Brunswick No. 12, the average tonnage reduces to 5.72 million tonnes. Average grade and tonnage of the 25 major Kuroko deposits of the Green Tuff Belt of Japan is 5.81 million tonnes containing 1.63% Cu, 3.86% Zn, 0.92% Pb, 12.17 g Ag/tonne and 0.37 g Au/tonne.

GEOLOGICAL SETTING

(9.1.a) Submarine, predominantly mafic volcanic sequences, e.g. greenstone belts, ophiolite sequences.
(9.1.b) Submarine volcanic sequences, commonly consisting of a bimodal mafic-felsic composition, situated on continental crust. Submarine sedimentary rocks usually constitute >40% of total succession.

HOST ROCKS OR MINERALIZED ROCKS

(9.1.a) Tholeiitic or calc-alkaline volcanic sequences. Deposits are commonly clustered around centres of felsic volcanism.
(9.1.b) Predominantly non-alkaline felsic volcanic rocks and argillaceous to arenaceous clastic sedimentary rocks.

ASSOCIATED ROCKS

Rhyolite domes; phreatic explosion breccias; volcanic derived sedimentary rocks including debris flows and greywacke; chemical sedimentary rocks including sulphidic chert, iron oxide, manganese oxide; carbonaceous argillites. Stratigraphic footwall and to some extent the stratigraphic hanging wall of those deposits with a contiguous stringer ore zone are hydrothermally altered, and are characterized by varied amounts of Mg, Si, K, Ca and Na metasomatism.

FORM OF DEPOSIT, DISTRIBUTION OF ORE MINERALS

Predominantly pyritic lensoid to tabular concordant massive sulphide bodies, usually stratigraphically underlain by highly variable amounts of discordant stringer sulphide ore and hydrothermally altered wallrock. In unmetamorphosed deposits, massive ore is usually very fine grained. Textures of lensoid or domical massive sulphide bodies are more typically massive, rubbly or brecciated, often surrounded by an apron of coarsely layered "clastic" massive sulphide. Tabular bodies are more typically conformably layered or laminated.
Ore minerals characteristically zoned concentrically outwards from the core of the stringer zone and upwards and outwards from the stringer/massive ore contact. In the case of
(9.1.a) cpy + po ± mag → py → sphal
(9.1.b) cpy + py ± po → sphal + gal → barite
Several Cu-Zn deposits (9.1.a) have a well-documented Mg-metasomatized zone (generally chloritic) coinciding with the core of the stringer sulphide ore, and grading outward to a more potassic alteration assemblage. Documented alteration at the Kuroko deposits, and observations at several other Zn-Pb-Cu deposits (9.1.b) by two of the authors (JMF, DFS) indicate a core zone of silicification and potassic alteration and a peripheral zone of chloritic alteration.

MINERALS: Principal ore minerals
– *Associated minerals*

(9.1.a) Sphalerite, chalcopyrite, galena.
– Pyrite, pyrrhotite, bornite, magnetite, sulphosalt minerals, native silver, cassiterite
(9.1.b) Sphalerite, galena, chalcopyrite.
– Pyrite, pyrrhotite, hematite, magnetite, sulphosalt minerals, barite, gypsum
– Hydrothermal silicate alteration products include quartz, chlorite, smectite, sericite, and iron, calcium and magnesium carbonates. In highly metamorphosed deposits, alteration assemblages are represented by cordierite, anthophyllite, biotite, talc, kyanite, sericite, garnet, staurolite, gahnite

AGE, HOST ROCKS

From 3.7 Ga to present day. In Canada most of the mining production is from deposits of Archean (2.65 - 2.75 Ga), Aphebian (1.7 - 1.9 Ga) or Cambro-Ordovician age.

AGE, ORE

Same as host rocks.

GENETIC MODEL

Deposition on or below the sea floor from solutions discharged from high temperature submarine hydrothermal systems. The discharge sites are fracture-controlled. Intrusion and extrusion of igneous melts, especially of felsic composition, often controlled by same fracture systems. The composition of the hydrothermal solutions, and hence the composition of the ore deposit, is controlled by the rock-forming mineral assemblage and temperature of the hydrothermal reservoir. Hence subtype 9.1.a is related to the ferromagnesian mineral-feldspar assemblages of mafic rocks and subtype 9.1.b is related to the feldspar-mica-clay mineral assemblages of sediments and/or felsic rocks. The relatively high copper content of volcanic associated massive sulphide deposits as compared to sediment-hosted deposits (subtype 9.2) is controlled by the higher temperature (>250°C approx.) of the hydrothermal solutions of the former. After initial deposition, the massive sulphide mounds may be mechanically transported downslope from the vent area.

ORE CONTROLS, GUIDES TO EXPLORATION

1. Along synvolcanic fractures in successions of submarine volcanic rocks.
2. Within or around felsic volcanic centres.
3. Within a given district, deposits tend to preferentially occur at a specific stratigraphic horizon.
4. Deposits tend to occur in clusters having diameters in the 20-40 km range.
5. Hydrothermal alteration of host rocks: a) intense magnesium metasomatism of immediately contiguous footwall rocks, b) silicification laterally along ore horizon, c) possibly widespread Na-depletion zone in stratigraphic footwall succession d) possibly Na addition in stratigraphic hanging wall immediately above deposit.
6. Pyroclastic volcanic rocks in immediate stratigraphic footwall.

AUTHORS

J.W. Lydon, J.M. Franklin, D.F. Sangster

(see Plates 6, 7, 8, pages 8, 9)

9.2 SEDIMENT-HOSTED SULPHIDE

COMMODITIES

Zn, Pb, Ag, barite (Cd, Cu, Sn)

EXAMPLES:
Canadian – *Foreign*

Sullivan, Cirque, B.C.; Faro, Howards Pass, Tom and Jason, Yukon; Walton, N.S. is probably a deposit of this type, comparable to Silvermines, Ireland. – *Balmat, New York; Broken Hill, Mt. Isa and McArthur River, Australia; Broken Hill and Gamsberg, South Africa; Rammelsberg and Meggen, West Germany; Silvermines and Tynagh, Ireland*

IMPORTANCE

Canada: in 1977-78, 16% of the zinc, 45% of the lead and 10% of the silver produced in Canada was from this type of deposit. These proportions will probably increase in the future.
World: currently, the bulk of the world's known reserves of zinc and lead in deposits of this type occur in Australia, Canada, and South Africa.

TYPICAL GRADE, TONNAGE

Range (and weighted average) of 38 worldwide examples: 4 to 550 (av. 60) million tonnes; 0.6% to 18% (av. 7.3%) Zn; 0.3% to 13% (av. 4.0%) Pb; nil to 1.0% (av. 0.1%) Cu; trace to 180 g/tonne (av. 48 g/tonne) Ag. Some deposits have large reserves of barite associated with the sulphide ores, e.g. Walton, N.S. (now closed) produced about 4 million tonnes $BaSO_4$. Meggen, Germany produced about 7 million tonnes $BaSO_4$. Anvil district deposits (e.g. Faro), Tom and Jason, Yukon and Cirque, B.C. have substantial barite contents.

GEOLOGICAL SETTING

Within second order, often tectonically (growth fault) controlled sedimentary basins situated in a continental rise, continental shelf or intracontinental marine basin.

HOST ROCKS OR MINERALIZED ROCKS

Deep marine clastic sedimentary rocks (shales, siltstones, fine to coarse grained turbidites), starved basin lithofacies (carbonaceous to siliceous shales, chert), shallow marine lithofacies (calcareous shales, carbonates).

ASSOCIATED ROCKS

Sedimentary breccias and conglomerates, especially in the stratigraphic footwall; talus from synsedimentary fault scarp. Sulphide zone may be overlain by, or pass laterally into, chemical sediments, particularly chert and baritite. Minor amounts of volcanic rocks, especially tuffs, recognized in host rocks of some deposits. Discordant feeder zone may be silicified, carbonatized, tourmalinized. Increase in biogenic activity near hydrothermal vents may be indicated by increase in carbon, silica and phosphorus content of associated rocks.

FORM OF DEPOSIT, DISTRIBUTION OF ORE MINERALS

Concordant interbedded layers of sulphide and host rocks form mineralized bodies whose lateral extents are tens to hundreds of times greater than their thicknesses. Ores are typically bedded on a scale varying from a few microns to several centimetres. Individual sulphide beds are often monomineralic. Relatively small "feeder zones" discordant to stratiform mineralization have been identified in many deposits. Pb/Zn, Cu/Zn, Zn/Ba ratios of the stratiform mineralization typically decrease away from the feeder zone.

MINERALS: Principal ore minerals
– *Associated minerals*

Sphalerite, galena, barite.
– *Quartz, pyrrhotite, pyrite, chalcopyrite, marcasite, arsenopyrite, sulphosalts, cassiterite*

AGE, HOST ROCKS

Canada: Sullivan, 1.43 Ga; Northern B.C.-Yukon, 0.55-0.34 Ga; Australia, South Africa, 2.0-1.7 Ga; Europe, 0.38-0.36 Ga.

AGE, ORE

Same as host rocks.

GENETIC MODEL

Deposition in a brine pool in a second order basin. Discharge temperature of fluids is generally less than that of volcanic-associated deposits (9.1), i.e., probably in the range 150-250°C. Hydrothermal activity is associated with tectonic activity, manifested by growth faults, slump breccias, etc. Some deposits may be a product of low heat flux discharge or seepage of stratafugic water (e.g., derived by compaction of underlying sedimentary pile) into a euxinic, starved basin environment.

ORE CONTROLS, GUIDES TO EXPLORATION

1. The majority of deposits are spatially associated with intracontinental or continental margin basins - usually thick successions of clastic sedimentary rocks.
2. Second order basins are prime exploration targets, and are recognized by local lithological facies that are additional or exotic to the regional lithological succession, and rapid lateral facies change.
3. Evidence of syndepositional tectonic activity: growth faults, fault scarp talus, slump and slide breccias. Tectonically active zone may represent reactivation of basement faults.
4. Evidence of syndepositional geothermal activity: presence of volcanic rocks in the succession, usually local flows or thin tuff horizons; presence of other chemical sediments of hydrothermal origin (e.g. chert, baritite, sediments enriched in iron and manganese), or of biogenic origin resulting from hydrothermal activity (e.g. sediments enriched in carbon, phosphorous, and silica).

AUTHORS

J.W. Lydon, D.F. Sangster

(see Plate 9, page 9)

9.3 SEDIMENT-HOSTED BARITE

COMMODITIES	Barite
EXAMPLES: Canadian – ***Foreign***	Cathy, Oro, and Tea deposits, Yukon; Sulphur Creek, B.C. – *Battle Mountain, Nevada*
IMPORTANCE	Canada: no production at present in Canada, but large potential in Ordovician to Mississippian basinal facies sedimentary rocks of northern British Columbia and Yukon Territory. World: this suite of rocks extends from the Arctic to Mexico, and contains the most important barite deposits in the United States, including the Battle Mountain area, from which 1.7 million tonnes or 23% of world barite production was derived in 1979.
TYPICAL GRADE, TONNAGE	One to more than 10 million tonnes of 30% to 90% $BaSO_4$. The Tea deposit is reported to contain more than 75 000 tonnes of direct shipping barite. The Cathy deposit is reported to have a much larger reserve of 50% $BaSO_4$.
GEOLOGICAL SETTING	Mainly epicontinental marine sedimentary basins.
HOST ROCKS OR MINERALIZED ROCKS	Marine basinal facies sedimentary rocks, particularly siliceous, carbonaceous and calcareous argillites to arenites.
ASSOCIATED ROCKS	Other chemical sedimentary rocks, especially chert and carbonate rocks.
FORM OF DEPOSIT, DISTRIBUTION OF ORE MINERALS	Concordant, lensoid to tabular bodies. Ores are typically massive or interbedded with chert, limestone or other host rock lithologies. Fine grain size and fine lamination parallel to bedding are common textural characteristics. Nodules or rosettes of barite commonly occur in lower grade beds.
MINERALS: Principal ore minerals – ***Associated minerals***	Barite. – *Quartz, calcite, and other common sedimentary rock-forming minerals; barium carbonate, barium aluminosilicates*
AGE, HOST ROCKS	Ordovician-Mississippian sedimentary rocks of western North America (from Arctic to Mexico) contain most of the known indicated reserves of barite. On a global scale, occurrences range in age from Archean (Fig Tree Group, South Africa >3.2 Ga) to modern ocean floor (e.g. Guayamas Basin).
AGE, ORE	Same as host rocks.
GENETIC MODEL	1. Most of the deposits having economic potential (i.e. higher grade) are localized, thick lenses. These lenses have been interpreted as resulting from barite deposition near low temperature (probably <100°C) springs or groundwater seepages, due to fixation of the barium content of springwater by seawater sulphate. 2. An alternative model for high grade beds or lenses is mechanical concentration of the heavy barite fraction by reworking or erosion/sorting of a low grade baritic source, as suggested for deposits of the Karoo Supergroup, South Africa. 3. It has also been suggested that some barite deposits were formed by oxygenation of barium-bearing seawater in euxinic marine basins by global climatic changes, resulting in synchronous barite deposition over large areas. This model is probably more applicable to the laterally extensive, often thin or low grade deposits.
ORE CONTROLS, GUIDES TO EXPLORATION	1.* Same as for sediment hosted stratiform sulphide deposits. In structurally controlled, second order sedimentary basins, especially along tectonic/structural lineaments marked by rapid lithological change, fault-scarp talus breccias, slump structures, etc. Presence of other chemical sediment, e.g. chert, sulphides, carbonates, is a positive indicator. 2.* Stratigraphically above erosional hiatus of baritic source rocks. 3.* At the same stratigraphic horizon and in similar lithologies to known barite deposits. Evidence of sudden change in chemical environment from reducing to oxidizing conditions is a positive indicator, e.g. change from carbonaceous to calcareous sediments; increase in biologic activity, such as marked by increase of phosphate content in sedimentary rocks.
AUTHORS	J.W. Lydon, K.R. Dawson

(see Plate 10, page 10)

* Refer to Genetic Model.

10. VOLCANIC REDBED COPPER

COMMODITIES

Cu (Ag)

EXAMPLES: Canadian – *Foreign*

Sustut, B.C.; White River, Yukon; Coppermine River area, N.W.T.; Mamainse Point, Ont. – *Keweenaw Peninsula, Michigan; Boléo, Mexico; Jardin and Buena Esperanza, Chile*

IMPORTANCE

Canada: no production at present but potentially a small percentage of Canadian copper production. World: was the most important copper producing type in USA from 1845-1885 (Keweenaw Peninsula). Generally of minor importance now.

TYPICAL GRADE, TONNAGE

0.6 to 4.0% Cu; 1 to 10 million tonnes. The Sustut deposit in B.C. is somewhat larger, and some of the Keweenaw Peninsula deposits contained more than 50 million tonnes of ore.

GEOLOGICAL SETTING

Continental to very shallow marine volcanic sequences deposited in low to possibly intermediate latitude, desert to semi-arid environments. Deposits occur mainly in continental rift-related flood basalt sequences, but are found also in island and continental arc sequences.

HOST ROCKS OR MINERALIZED ROCKS

Amygdaloidal flows; mafic to felsic tuff and breccia; and interlayered, locally-derived, carbonaceous, clastic sedimentary rocks (conglomerate, sandstone and siltstone).

ASSOCIATED ROCKS

Various volcanic rocks, especially amygdaloidal basalt, and interlayered sedimentary rocks. Reddish, oxidized varieties characteristically present in the sequence. Bedded evaporites and thin marine limestones are also present in some areas.

FORM OF DEPOSIT, DISTRIBUTION OF ORE MINERALS

Highly variable. The largest deposits tend to be concordant or peneconcordant and to follow specific lithologies such as amygdaloidal flow top breccia; pyroclastic tuff and breccia; and interlayered conglomerate, carbonaceous sandstone and siltstone. Many smaller deposits occur as veins or irregular stringer zones in fissures and faults and fault breccias. In some deposits, concordant zones extend short distances away from the cross-cutting stringer zone. Ore minerals occur as disseminations, stringers, lenses, and irregular, patchy accumulations.

MINERALS: Principal ore minerals – *Associated minerals*

Native copper, chalcocite, bornite, chalcopyrite, other copper minerals, native silver.
– *Calcite, quartz, epidote, prehnite, chlorite, pumpellyite, laumontite, K-feldspar, albite*

AGE, HOST ROCKS

Similar to sedimentary copper deposits - Early Proterozoic (about 2.25 Ga) to Tertiary. Common in Late Proterozoic (1.0 to 1.4 Ga) and Late Triassic-Early Jurassic rocks in North America.

AGE, ORE

Penecontemporaneous with, or slightly younger than, host rocks.

GENETIC MODEL

Controversial, poorly understood. The model generally proposed involves mobilization of copper during low-grade regional metamorphism from volcanic rocks, transport upward and outward along the metamorphic gradient, and deposition under lower temperature and pressure conditions.
An alternative, though less widely accepted model (cf. subtype 6.3 Sedimentary Copper) invokes mobilization of copper out of volcanic rocks during diagenesis or early metamorphism (possibly related to synvolcanic geothermal activity), transport through oxidized continental volcanic facies, and deposition on encountering reducing environments.

ORE CONTROLS, GUIDES TO EXPLORATION

1. Permeable zones:
   - stratigraphic units, such as flow top breccias, tuffs, volcanic breccias, interlayered conglomerates and sandstones.
   - fault and fracture systems, zones of extensive fault brecciation.
2. Favourable host rocks such as anoxic carbonaceous pyritic units in volcanic redbed sequences.

AUTHOR

R.V. Kirkham

(see Plate 11, page 10)

11. VOLCANIC—ASSOCIATED VEIN AND SHEAR ZONE GOLD

COMMODITIES
Au (Ag)

EXAMPLES:
Canadian – *Foreign*
Con and Giant Yellowknife Mines, N.W.T.; Dome and Pamour Mines at Timmins, Campbell Mine at Red Lake, and Kerr Addison Mine at Virginiatown, Ontario – *Norseman, Australia*

IMPORTANCE
Canada: estimated to account for about 25% of Canada's total cumulative gold production. World: of major importance in Zimbabwe and Australia, as well as Canada.

TYPICAL GRADE, TONNAGE
Up to 40 million tonnes averaging 8.6 g Au/tonne. Many long term producers range from 1 to 6 million tonnes at grades of 7 g Au/tonne.

GEOLOGICAL SETTING
Archean greenstone belt terrane. Many are associated with highly schistose zones of intense rock alteration ("shear zones") interpreted as major structural breaks; and are in areas that also contain intrusion-associated gold deposits (e.g. Timmins, Red Lake). Veins occur mostly in greenschist metamorphic domains.

HOST ROCKS OR MINERALIZED ROCKS
Tholeiitic pillowed basalts; thin units of komatiitic volcanic rocks; pyroclastic rocks of either of these types; greywacke and conglomerate; highly altered rocks (carbonatized, sericitized, pyritized, less commonly albitized).

ASSOCIATED ROCKS
Volcaniclastic rocks; greywacke and conglomerate of predominantly greenstone belt provenance; less commonly iron formation, other chemical sedimentary rocks, felsic porphyritic intrusive bodies, and mafic or ultramafic intrusive rocks.

FORM OF DEPOSIT, DISTRIBUTION OF ORE MINERALS
Veins and irregular bodies of quartz along fractures and faults, or along zones of slightly to highly altered and schistose rock. Gold is associated with sulphides, disseminated, or as small irregular patches, in quartz.

MINERALS: Principal ore minerals
– *Associated minerals*
Native gold, tellurides.
– Pyrite, arsenopyrite, minor amounts of other sulphides; quartz, minor sericite, carbonates. Fuchsite is present in many major deposits

AGE, HOST ROCKS
Cited examples are all Archean. There are probably Mesozoic analogues, e.g., Carolin and Bralorne, B.C., and Motherlode, California.

AGE, ORE
Archean (the available evidence favours an age not much younger than host rock age).

GENETIC MODEL
Three models that have been suggested are (1) lateral secretion from the host rocks (Boyle, 1961) and (2) deposition from large volumes of metamorphically generated fluids (Kerrich and Fryer, 1979), and (3) syngenetic exhalative accumulation (Karvinen, 1981).
We prefer a model in which gold and associated elements are scavenged from large volumes of rocks, possibly by fluids derived from the greenschist-amphibolite metamorphic transition, and deposited in fractures, faults and shear zones. Associated hydrothermal alteration, commonly including carbonatization, is in some cases extensive and pervasive, possibly indicating long-lived, CO_2-rich, major hydrothermal systems.

ORE CONTROLS, GUIDES TO EXPLORATION
1. Archean age.
2. Fault or "shear zone" structures.
3. Presence of thin komatiitic units, highly altered (especially carbonatized) rocks, and sulphide-bearing quartz veins.
4. Fuchsite may be a guide in some cases.

AUTHORS
R.I. Thorpe, J.M. Franklin

Figure 12. *11 Volcanic-Associated Vein and Shear Zone Gold. Kerr-Addison mine, Virginiatown, Ontario. Typical, irregular gold-bearing quartz veins of the green carbonate zone. Width of field of view is about 1.5 m. Photo: R.V. Kirkham (GSC 203633-S).*

12. MAGMATIC NICKEL, COPPER, PLATINUM GROUP ELEMENTS

12.1 ULTRAMAFIC-ASSOCIATED NICKEL, COPPER

12.1.a Volcanic Peridotite Nickel
12.1.b Intrusive Dunite Nickel
12.1.c Intrusive Ultramafic Nickel-Copper

COMMODITIES

Ni (Cu, Pt, Pd, Co)

EXAMPLES: Canadian – *Foreign*

(12.1.a) Langmuir, Ont.; Marbridge, Que. – *the Kambalda group of mines, Nepean, Scotia, and Windarra, Australia; Shangani, Zimbabwe*
(12.1.b) Pipe, Birchtree and Manibridge mines, Thompson, Man.; Dumont, Que. – *Mt. Keith, and Agnew, Australia*
(12.1.c) Shebandowan, Ont.; Ungava deposits, Que. – *Pechenga deposits, U.S.S.R.*

IMPORTANCE

Canada: the Thompson, Man. deposits, probably of this type (12.1.b), accounted for 15-20% of Canadian annual nickel production during the late 1970s. Current and past producers of subtypes 12.1.a and 12.1.c accounted for about 5%.
World: deposit types 12.1.a and 12.1.b account for virtually all current and past nickel production in Western Australia and Zimbabwe.

TYPICAL GRADE, TONNAGE

(12.1.a) From less than 1 million up to 5 million tonnes per ore body, commonly several ore bodies per deposit. 1 to 5% Ni, 0.1 to 0.25% Cu; Ni/Cu = 10 to 20.
(12.1.b) Segregated ores (see Form of Deposit) comprise millions and tens of millions of tonnes at 1 to 3% Ni, 0.04 to 0.2% Cu; Ni/Cu = 10 to 20. Disseminated ores, up to 250 million tonnes at about 0.6% Ni, 0.01% Cu; Ni/Cu>20.
(12.1.c) Commonly less than 1 million to a few million tonnes. Some probably greater than 10 million tonnes. 1 to 3% Ni, 0.4 to 1% Cu; Ni/Cu = 1.5 to 4.

GEOLOGICAL SETTING

(12.1.a) Archean greenstone belts.
(12.1.b) Thompson deposits occur in an Aphebian supracrustal fold belt marginal to an Archean craton. Others are mainly in Archean greenstone belts.
(12.1.c) Proterozoic mafic volcanic belts, Archean greenstone belts.

HOST ROCKS OR MINERALIZED ROCKS

(12.1.a) Komatiites (highly magnesian ultramafic flows and closely underlying small sills), serpentinized, commonly carbonatized, in some cases adjacent metasedimentary rocks, metavolcanic rocks.
(12.1.b) Highly magnesian ultramafic sills, generally larger than those in subtype 12.1.a.
(12.1.c) Ultramafic intrusive lenses, less magnesian than in subtypes 12.1.a and 12.1.b; adjacent metasedimentary and metavolcanic rocks.

ASSOCIATED ROCKS

Komatiitic basalts, tholeiitic basalts, felsic pyroclastics, slates, iron formation (commonly sulphide, less commonly oxide).

FORM OF DEPOSIT, DISTRIBUTION OF ORE MINERALS

(12.1.a, 12.1.c) Commonly basal, conformable, tabular or lens-shaped bodies within the ultramafic host. Ore types include massive, intercumulus (or matrix), and disseminated sulphides; this sequence is the one observed in successive upward zones beginning at the basal contact in some well preserved deposits.
(12.1.b) Ores may comprise either 1) rich segregations at margins (basal contacts where interpretable) of ultramafic lenses, as at Agnew, Pipe and Manibridge; or 2) conformable internal zones of disseminated sulphides, as at Dumont and Mt. Keith.
Remobilization of sulphides into veins, breccia matrix, and disseminations in fault zones and wall rocks is common.

MINERALS: Principal ore minerals – *Associated minerals*

Pentlandite, chalcopyrite; in some cases millerite, violarite, bornite.
– *Pyrrhotite, pyrite, magnetite, various platinum group minerals, serpentine, tremolite, clinopyroxene, chromite, magnesite, chlorite, rarely relict olivine*

AGE, HOST ROCKS

(12.1.a) Mainly Archean, 2.9 to 2.7 Ga.
(12.1.b) Mainly Archean, 2.9 to 2.7 Ga. Thompson deposits probably Aphebian.
(12.1.c) Archean, 2.9 to 2.7 Ga; lower to mid-Proterozoic.

AGE, ORE

Same as ultramafic host rocks.

GENETIC MODEL

Partial melting of the mantle is considered to have produced highly magnesian liquids (>20% MgO in Archean komatiitic host rocks subtypes 12.1.a and 12.1.b). In the case of deposits containing rich basal concentrations of nickel sulphides, the liquid apparently became saturated with respect to sulphur prior to, or at an early stage of, crystallization. The resulting immiscible nickeliferous sulphide droplets became segregated by flow and gravitational settling, and gave rise to rich basal sulphide concentrations in the ultramafic flows and sills. In the case of deposits consisting of internal zones of disseminated sulphides, sulphur saturation was apparently reached at a later stage of crystallization, probably in situ within the ultramafic flows and sills.
The apparent high degree of mantle melting that gave rise to komatiitic rocks and their associated nickel ores seems to have been most common in Archean times, but also occurred in the Aphebian (Thompson). Subtype 12.1.c is presumed to be of similar genesis but involved liquids that were less magnesian.

The decrease of MgO in parent liquids and corresponding decrease of Ni/Cu ratio in associated ores is attributed to a combination of 1) diminishing degrees of partial melting of the mantle, and 2) progressive degrees of fractional crystallization of the parental magma. Sulphur in the ores may have been of either mantle or crustal origin.

ORE CONTROLS, GUIDES TO EXPLORATION

(12.1.a) Komatiitic flows and sills, commonly the lowest and most magnesian komatiites in the ultramafic pile; also commonly closely associated with sulphide facies iron formation near the top of the underlying felsic volcanic-sedimentary cycle. Basal portions of komatiitic bodies, especially in original depressions along basal contacts.

(12.1.b) Komatiitic sills, weakly to strongly differentiated; sulphides tend to occur in internal layers and basal depressions.

(12.1.c) Basal portions of weakly differentiated ultramafic intrusive lenses.

AUTHOR

O.R. Eckstrand

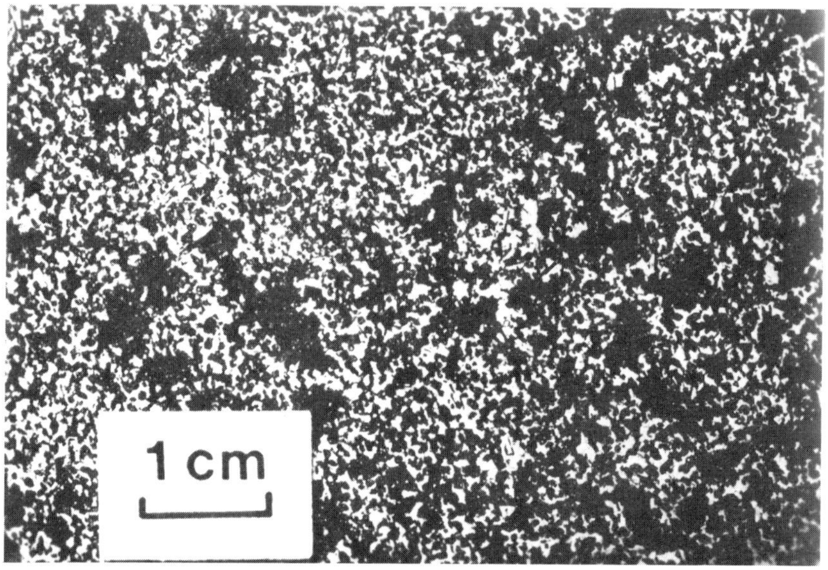

Figure 13. *12.1.a Volcanic Peridotite Nickel. Alexo mine, near Timmins, Ontario. Typical intercumulus or matrix nickel sulphide ore in an ultramafic host rock. Pyrrhotite and pentlandite (white) form a continuous interstitial network in serpentinized peridotite consisting mainly of equant serpentine pseudomorphs (black) after olivine. This texture is generally interpreted as a primary magmatic feature. Photo: O.R. Eckstrand (GSC 203887).*

Figure 14. *12.1.c Intrusive Ultramafic Nickel-Copper. Shebandowan mine, Thunder Bay, Ontario. Breccia ore. A fine grained matrix of sulphides (pyrrhotite, pentlandite, chalcopyrite) and magnetite enclose fragments of mafic metavolcanic wall rocks in a texture typical of many deformed nickel sulphide deposits. Sample: O.R. Eckstrand (GSC 203886-F).*

12.2 GABBROID-ASSOCIATED NICKEL, COPPER, PLATINUM GROUP ELEMENTS

12.2.a Layered Intrusive, Nickel-Copper
12.2.b Layered Intrusive, Platinum Group Elements
12.2.c Stock

COMMODITIES

Ni, Cu, Platinum Group Elements (Co, Au, Ag, S, Fe)

EXAMPLES:
Canadian – *Foreign*

(12.2.a) Sudbury deposits, Great Lakes Nickel, Ont. – *Duluth Complex, Minnesota; Stillwater Complex nickel-copper deposits, Montana; Brady Glacier, Alaska; Noril'sk-Talnakh deposits, U.S.S.R.; Pikwe and Selebi deposits, Botswana; Insizwa-Ingeli, (southern Africa) Transkei*

(12.2.b) Lac des Isles, Ont. – *Stillwater Complex platinum deposits, Montana; Merensky Reef, Bushveld Complex, South Africa*

(12.2.c) Lynn Lake, Man.; Giant Mascot, B.C.; St. Stephen, N.B. – *Carr Boyd, Australia*

IMPORTANCE

(12.2.a) Canada: the Sudbury deposits have, by a considerable margin, produced more nickel (about 7 million tonnes) than any other district in the world, as well as substantial copper (about 6 million tonnes), precious metals and other byproducts.
World: this type is estimated to account for about 80% of the world's reserves of sulphide nickel, and about one half of current world production of Platinum Group Elements (PGE).

(12.2.b) Canada: no producers.
World: Merensky Reef accounted for about one half of world PGE production, 1977-79.

(12.2.c) Canada: the Lynn Lake deposits (largest example in Canada by an order of magnitude) produced about 200 000 tonnes of nickel and 95 000 tonnes of copper, as well as some cobalt and PGE.
World: few producers; of relatively minor importance.

TYPICAL GRADE, TONNAGE

Individual bodies may contain from a few hundred thousand tonnes to tens of millions of tonnes of ore, and each intrusive complex generally contains a number of ore bodies. Grades generally range from about 0.6 to 1.6% Ni, 0.2 to 1.3% Cu but large, lower grade deposits are known (e.g. Great Lakes Nickel, 0.20% Ni, 0.36% Cu). Combined PGE content (Pd and Pt mainly) is generally of the order of one g/tonne, but Stillwater, Merensky Reef and Talnakh deposits contain PGE in the 10-20 g/tonne range.
Examples: Sudbury (production + reserves): 700 million tonnes, 1.6% Ni, 1.3% Cu, in numerous deposits clustered in 5 main areas. Lynn Lake (production): 20 million tonnes, 1.02% Ni, 0.54% Cu, in 11 ore bodies.

GEOLOGICAL SETTING

(12.2.a), (12.2.b) Layered intrusions generally occur in a cratonic setting, in some cases associated with intracontinental rifts and flood basalts. Some occur in Archean greenstone belts. Sudbury deposits are related to a structure of probable meteoritic impact origin.

(12.2.c) The stock-like intrusions occur in Precambrian greenstone belts and younger orogenic belts.

HOST ROCKS OR MINERALIZED ROCKS

Various mafic phases of intrusive complexes; includes norite, gabbro, troctolite, feldspathic pyroxenite, amphibolite, gabbro-diabase, picrite. The North Range deposits at Sudbury occur in brecciated leucocratic footwall rocks. The Merensky Reef is coarse grained feldspathic pyroxenite with associated thin chromitite layers.

ASSOCIATED ROCKS

A variety of phases of the mafic intrusive complexes; includes diorite, peridotite, pyroxenite, anorthosite, gabbro, norite; wall rocks of the intrusive complexes.

FORM OF DEPOSIT, DISTRIBUTION OF ORE MINERALS

(12.2.a) Conformable layers or lenses, commonly located in a local depression or embayment at or near the base of the host layered intrusion. Ore consists of massive sulphides, sulphide-matrix breccia, interstitial sulphide network, and disseminated sulphide. In well preserved deposits, the rich ores lie nearest the base, and are overlain by leaner disseminated sulphide. Sulphide veins and disseminations commonly penetrate footwall rocks. At Sudbury, some of the deposits (North Range) occur as sulphide impregnations in leucocratic footwall breccias, and some occur as sulphide disseminations and breccia matrix in long, dyke-like gabbroic apophyses ("offsets") in footwall rocks.

(12.2.b) The PGE ores (e.g. Bushveld, Stillwater) form thin stratiform layers within the host intrusions, and are associated with cumulate layers of anorthosite, norite and bronzitite.

(12.2.c) Ores form irregular zones, in some cases pipe-like, within the host stocks. Ore consists of massive sulphide, sulphide-matrix breccia, disseminated sulphides and sulphide veins.

MINERALS: Principal ore minerals
– *Associated minerals*

Pentlandite, chalcopyrite, cubanite, millerite; various PGE minerals including sulphides, tellurides, arsenides and alloys.
– *Pyrrhotite, pyrite, sphalerite, millerite, marcasite; plagioclase, hypersthene, augite, olivine, hornblende, biotite, quartz, and a variety of alteration minerals*

AGE, HOST ROCKS

Various ages; most are Precambrian (Sudbury, 1.85 Ga; Great Lakes Nickel and Duluth Complex, 1.1 Ga; Bushveld Complex, 2.1 Ga; Stillwater Complex, 2.7 Ga; Lynn Lake $\geq$ 1.8 Ga) but Noril'sk-Talnakh intrusions are Permo-Triassic, and other Paleozoic and Mesozoic examples are known. Intrusions (subtype 12.2.c) may apparently be either syn- or post-orogenic.

AGE, ORE

Syngenetic with the host intrusions.

GENETIC MODEL

(12.2.a) Mafic magma (probably mantle-derived in most cases) was generally emplaced quiescently as multiple pulses in upper levels of the crust, in some cases apparently in a tensional environment associated with rifting. Early sulphur saturation of the magma produced flow- and gravity-segregations of Ni-Cu-bearing sulphides at the base of the intrusion. Contamination of the magma probably contributed importantly to sulphide saturation in many deposits either through addition of sulphur, or assimilation of siliceous material. Most of the sulphur in several large districts (Noril'sk-Talnakh, Duluth Complex) was probably derived from underlying sedimentary rocks.

(12.2.b) Magmas that produced PGE ores apparently reached sulphide-saturation at a much later, and perhaps crucial, stage in the crystallization of their layered intrusive hosts.

(12.2.c) Much less is known about the genesis of these ores, but it seems highly probable that they represent immiscible sulphides present in the magma at the time of emplacement.

ORE CONTROLS, GUIDES TO EXPLORATION

(12.2.a) The basal contacts (particularly embayments in the basal contacts) and immediately overlying zones (up to about 200 m thick) in layered intrusions are the most common sites of nickel-copper sulphide ores.

(12.2.b) PGE-rich zones in layered intrusions tend to occur as thin, sparsely sulphide-bearing layers at some appreciable height above the base of the intrusion.

(12.2.c) Differentiated, multiple phase stock-like intrusions.

AUTHOR

O.R. Eckstrand

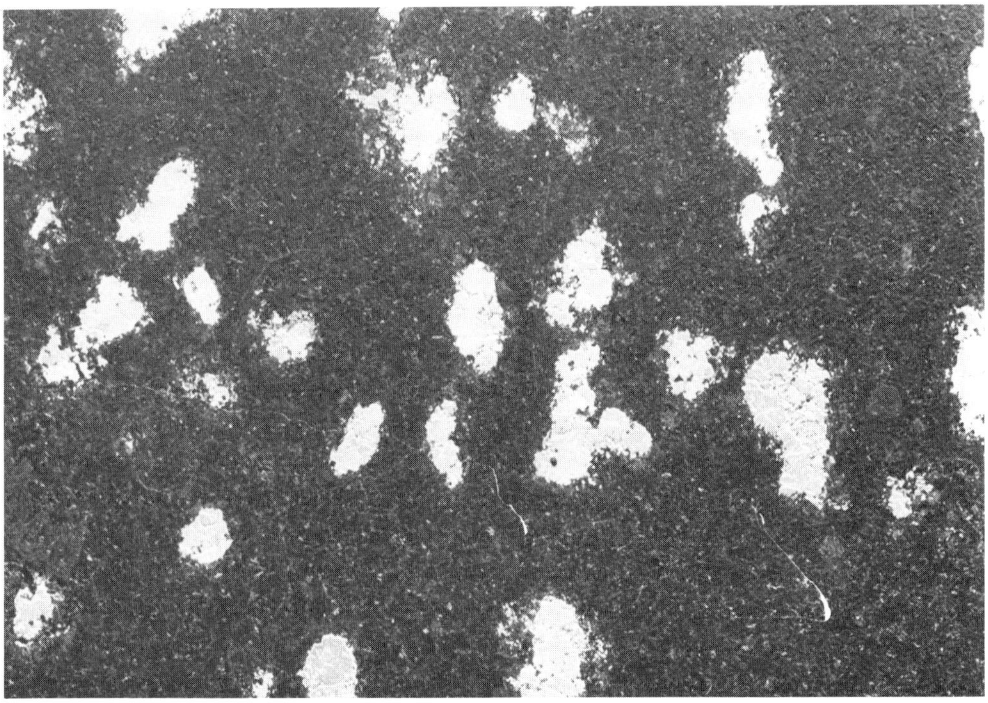

Figure 15. *12.2.a Layered Intrusive, Nickel-Copper. Clarabelle mine, Sudbury, Ontario. Disseminated nickel-copper sulphide ore. Blebs of pyrrhotite, pentlandite, and chalcopyrite in dark norite. Width of field of view is about 6 cm. Sample: O.R. Eckstrand (GSC 203886-V).*

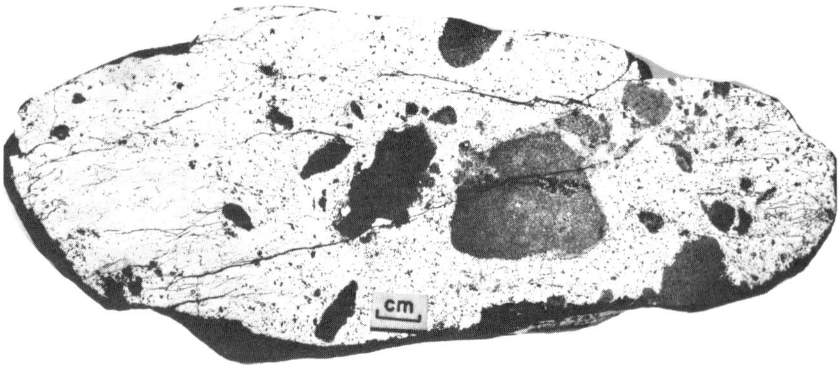

Figure 16. *12.2.a Layered Intrusive, Nickel-Copper. Murray mine, Sudbury, Ontario. Massive nickel-copper sulphide ore. The sulphide (white) consists mainly of a uniform granular mosaic of pyrrhotite with interstitial pentlandite and sparse irregular patches of chalcopyrite. Dark xenolithic fragments consist of fine grained mafic wall rocks. Sample: O.R. Eckstrand (GSC 203886-B).*

13. MAFIC/ULTRAMAFIC—HOSTED CHROMITE

13.1 STRATIFORM

COMMODITIES
Cr, Chromite

EXAMPLES: Canadian – ***Foreign***
Bird River Sill, Man.; Puddy Lake-Chrome Lake, Shebandowan, Crystal Lake Gabbro (Great Lakes Nickel), and Big Trout Lake, Ont.; Muskox Intrusion, N.W.T.; Lac des Montagnes, Que. – *Bushveld Complex, South Africa; Great Dyke, Zimbabwe; Kemi, Finland; Campo Formoso and Jacurici Valley, Brazil; Stillwater Complex, Montana*

IMPORTANCE
Canada: no production at present. Of all the known chromite deposits in Canada, those within the Bird River Sill are perhaps the most likely candidates for future development.
World: Stratiform deposits account for 45% of total world chromite production and 95% of reserves. The Bushveld Complex alone accounts for 35% of production, whereas production from the Great Dyke, Kemi, and the Brazilian deposits together contribute nearly all of the remaining 10%.

TYPICAL GRADE, TONNAGE
Canada: four deposits within the Bird River Sill range in grade from 4.6% to 25.2% Cr_2O_3, and from 600 000 to 10 million tonnes. Total "possible" reserves have been estimated conservatively as 19.4 million tonnes with a weighted average grade of 10.7%. The Cr/Fe ratio is 1.6 in the higher grade material.
Foreign: Bushveld reserves exceed 1100 million tonnes with average grades of 44% Cr_2O_3 and Cr/Fe = 1.5. Additional resources are probably five times reserves. The reserves of the Great Dyke have been estimated to exceed 600 million tonnes of which about 90% is of the high-Cr type with more than 46% Cr_2O_3 and Cr/Fe > 2. The ore reserves at Kemi are 50 million tonnes averaging 27% Cr_2O_3 with Cr/Fe = 1.55.

GEOLOGICAL SETTING
Deposits occur in large, layered intrusions which are commonly differentiated into a lower ultramafic zone and an upper mafic zone. The intrusions occur in a variety of tectonic settings. The Bushveld, Great Dyke, and Muskox intrusions are funnel-shaped bodies which are essentially unmetamorphosed and appear to postdate the main deformation of their host rocks. The Bird River and Big Trout Lake bodies are pre-kinematic and probably synvolcanic intrusions in Archean greenstone belt settings. The Kemi and Campo Formoso deposits occur in tabular intrusions at the unconformable contact between Archean granitic basement and overlying, mainly sedimentary, Proterozoic rocks. The Crystal Lake Gabbro was emplaced during flood basalt magmatism related to continental rifting.

HOST ROCKS OR MINERALIZED ROCKS
Chromitite seams are most commonly hosted by peridotite or its serpentinized equivalent. The Bushveld chromitites are interlayered with orthopyroxenite, anorthosite and norite. The chromitite layers in the Muskox Intrusion occur at the transition between peridotite and orthopyroxenite. The chromitite in the Crystal Lake Gabbro is hosted in anorthositic gabbro.

ASSOCIATED ROCKS
Dunite, pyroxenite, norite and gabbro.

FORM OF DEPOSIT, DISTRIBUTION OF ORE MINERALS
Chromite occurs in semi-massive to massive seams or layers ranging from less than 1 cm to more than 1 m in thickness. In the remarkable Kemi deposit, chromite achieves a thickness of up to 90 m. Chromite is also generally disseminated in the host rocks. There are typically many parallel chromitite layers in a given intrusion and the individual layers have remarkable lateral continuity.

MINERALS: Principal ore minerals – ***Associated minerals***
Chromite.
– *Olivine, pyroxene, biotite, serpentine, chlorite, tremolite, plagioclase, talc*

AGE, HOST ROCKS
All major deposits occur in Precambrian intrusions. The Great Dyke, Stillwater, Bird River, Puddy Lake - Chrome Lake, and Lac des Montagnes bodies are Archean. The Bushveld Complex is Aphebian, and the Muskox and Crystal Lake Gabbro intrusions are Helikian.

AGE, ORE
The chromite is syngenetic with its host intrusion.

GENETIC MODEL
The stratiform chromite deposits have clearly formed by magmatic segregation during fractional crystallization of mafic magma. In most cases magma may be described as generally basaltic in composition; however, in the case of the Great Dyke, there is some evidence to suggest a rather more ultramafic parent magma, thereby causing chromite to precipitate. The precise reasons why massive chromite cumulate layers form are not entirely understood. Irvine (1975, 1977) has suggested a mechanism whereby a chromite-saturated picritic tholeiite liquid becomes more siliceous by contamination (assimilation) with granitic material or alternatively by blending with a more siliceous differentiate of the parent magma, thereby causing chromite to precipitate.

ORE CONTROLS, GUIDES TO EXPLORATION
1. Stratiform chromite deposits occur in differentiated sill-like or funnel-shaped, mafic-ultramafic intrusions.
2. Chromitite seams are themselves igneous layers and so the presence of layering is a positive indication that the necessary igneous processes took place in the intrusion.
3. Chromitite seams are most commonly associated with peridotite although pyroxenite and anorthosite are also host rocks in some intrusions.
4. Accessory (disseminated?) chromite is often conspicuous in the host rocks of the seams.

AUTHOR
J.M. Duke

Figure 17. *13.1 Stratiform Mafic/Ultramafic-Hosted Chromite. Bird River area, eastern Manitoba. Stratiform chromitite layers (dark) separated by olivine cumulate layers (white, light grey); occurs in the ultramafic zone of the Bird River Sill. Photo: R.F.J. Scoates.*

13.2 PODIFORM

COMMODITIES

Cr, Chromite

EXAMPLES:
Canadian – ***Foreign***

Sterrett Mine, Reed-Belanger (Chromeraine) Mine, Caribou Mine and Montreal Pit, Eastern Townships area, Que.; Bluff Head Mine, Bay of Islands Complex, Nfld.; Scottie Creek, Ashcroft area, B.C. – *Kempirsai Massif, U.S.S.R.; Masinloc and Acoje Mines, Zambales Massif, Philippines; Kavak mine, Turkey*

IMPORTANCE

Canada: no production at the present time. About 250 000 tonnes of chromite were mined from podiform ores from 1894 to 1949, mainly in Quebec but also to a very limited extent in Newfoundland and British Columbia.
World: podiform deposits currently account for 55% of world chromite production but only about 5% of ore reserves. The most important producing countries are the U.S.S.R., Albania, Turkey, India and the Philippines.

TYPICAL GRADE, TONNAGE

Individual podiform chromite orebodies range from a few tens to a few millions of tonnes. The Coto orebody of the Masinloc Mine in the Philippines is probably the largest known having an estimated 13 million tonnes of high-Al (refractory grade) ore with 36.5% Cr_2O_3, 31% Al_2O_3 and Cr/Fe = 2.2. Most important mines exploit a number of pods. The Kavak Mine in Turkey includes 21 orebodies totalling about 2 million tonnes grading 28 to 30% Cr_2O_3 from which a concentrate with 51% Cr_2O_3 and Cr/Fe = 3.2 is produced. The "Twenty Years of Kazakh S.S.R." deposit in the U.S.S.R. is reported to contain in excess of 23 million tonnes grading better than 50% Cr_2O_3. The reserves plus past production of some Canadian deposits total 1.1 million tonnes grading 7 to 14% Cr_2O_3 for the Reed-Belanger deposit, about 180 000 tonnes grading 18% Cr_2O_3 for the Sterrett Mine, and about 60 000 tonnes grading 27% Cr_2O_3 for the Caribou Mine. Concentrates from the Canadian ores typically graded about 48% Cr_2O_3 with Cr/Fe = 2.4 to 2.8.

GEOLOGICAL SETTING

Podiform chromite deposits occur within the ultramafic parts of ophiolite successions (cf. "alpine peridotites"). The ore deposits may be hosted by either cumulate or tectonite ultramafic rocks but the latter are the more important on a worldwide basis.

HOST ROCKS OR MINERALIZED ROCKS

The immediate host rocks of the chromite pods are most commonly dunite or serpentinized dunite. Troctolite is the host rock in a few areas.

ASSOCIATED ROCKS

Harzburgite or its serpentinized equivalent.

FORM OF DEPOSIT, DISTRIBUTION OF ORE MINERALS

Podiform chromite deposits are generally lenticular bodies of massive to heavily disseminated chromite. Tabular, rod-shaped, and irregular bodies are also observed. Nodular textures, foliation, and banding are common. A given deposit usually comprises a number of individual pods. For example, in the main ore field of the Kempirsai Massif in the U.S.S.R. there are 28 deposits, each of which comprises up to 99 discrete orebodies.

MINERALS: Principal ore minerals
– ***Associated minerals***

Chromite.
– *Olivine, pyroxene, serpentine, magnetite, clinopyroxene, plagioclase*

AGE, HOST ROCKS

Podiform chromite deposits occur in Phanerozoic rocks. Those in the Canadian Appalachians are Cambro-Ordovician in age, whereas those in the Cordillera are mainly Permo-Triassic.

AGE, ORE

The ores are syngenetic with the host rocks.

GENETIC MODEL

Podiform chromite ores are magmatic segregation deposits which may form either during fractional crystallization of basaltic magma in the oceanic crust (those in the cumulate ultramafic rocks) or upper mantle, or during partial melting of mantle peridotite (those in the ultramafic tectonites). Many of the morphological features of the deposits result from deformation.

ORE CONTROLS, GUIDES TO EXPLORATION

The chromite pods most commonly occur within an envelope of dunite which in turn occurs in harzburgite country rocks, and are only rarely associated with rocks containing essential clinopyroxene. There is a tendency for the pods that occur in the tectonites to be near the contact with the overlying cumulates and, in some cases, in close proximity to the gabbro cumulates. The chromite pods tend to be localized in linear zones, in some cases in en echelon fashion.

AUTHOR

J.M. Duke

14. MAFIC INTRUSION—HOSTED TITANIUM—IRON

14.a Anorthosite-hosted Ilmenite
14.b Gabbroic-Anorthosite-hosted Titaniferous Magnetite

COMMODITIES

(14.a) Ti, Fe (V)
(14.b) Fe (Ti, V, Cr, P)

* "Titaniferous magnetite" is used here in the common commercial sense, referring to magnetite-rich, Ti-bearing oxide ore. It generally consists of Ti-bearing magnetite and ilmenite occurring together as granular aggregates and exsolution intergrowths.

EXAMPLES:
Canadian – *Foreign*

(14.a) Lac Allard, St. Urbain, and Ivry, Que. – *Tellnes and Egersund, Norway; Ilmen Mountains, U.S.S.R.*
(14b) Magpie Mountain, Saint Charles, and Lac Doré Complex, Que.; Newboro Lake, Ont. – *Smaalands-Taberg, Sweden; Bushveld Complex, South Africa; Kusinskoye, Ural Mountains, U.S.S.R.; Tahawus, New York; Iron Mountain, Wyoming*

IMPORTANCE

(14.a) Deposits of this type account for most of Canadian, and more than one half of the world's production of TiO_2. Many undeveloped deposits are potential sources of titanium and in some cases of vanadium and coproduct iron.
(14.b) Canada: a few minor past producers, no current producers. Large potential resources. World: substantial current production from deposits in U.S.S.R., U.S.A.

TYPICAL GRADE, TONNAGE

(14.a) 1 to 300 million tonnes; TiO_2 from 10 to 45%; Fe from 32 to 45%; V about 0.2% or less. Fe:Ti ratio commonly about 2. Cu, Cr, Mn, and Ni each commonly range from 0.05 to 0.2% in variable relative proportions; sulphur and phosphorus contents low and variable.
(14.b) Less than 1 million tonnes to greater than 1000 million tonnes. Iron content commonly ranges from 20% to 45%; TiO_2 from 2 to 20%; Fe:Ti ratios range from 40:1 to 2:1 and commonly are about 5:1; V about 0.25%; traces of Cr; P_2O_5 variable, reported up to 7.1%.

GEOLOGICAL SETTING

Deposits occur mainly in differentiated gabbroic (noritic) anothosite intrusions. Most abundant in Precambrian Shield areas. In the Grenville province, these intrusions are set mainly in gneissic and granitic terrane.

HOST ROCKS OR MINERALIZED ROCKS

(14.a) Anorthosite, noritic anorthosite, andesine-anorthosite. Some host rocks are part of layered stratiform intrusions, others are stock-like intrusions.
(14.b) Most host rocks are gabbroic and anorthositic-gabbroic phases of differentiated mafic intrusions; diabase, gabbroic diorite, quartz monzonite.

ASSOCIATED ROCKS

Comagmatic rocks include anorthosite, gabbro, pyroxene diorite, diabase, and probably monzonite, syenite, pegmatite and granite. Wall rocks to the intrusions include high grade metamorphic gneisses, schists, amphibolites, quartzites, and contact skarns.

FORM OF DEPOSIT, DISTRIBUTION OF ORE MINERALS

The oxide ores occur as:

- massive lenses, dykes, sills and irregular intrusions, and
- stratiform or irregular bodies consisting of disseminated and interstitial Fe-Ti oxide minerals in anorthositic (14.a) and gabbroic (14.b) host rocks. Both forms of mineralization are generally present, in various proportions in different deposits. The intrusive, massive form appears more important economically than the disseminated form.

MINERALS: Principal ore minerals
– *Associated minerals*

(14.a) Ferrian ilmenite, titanhematite and their exsolution intergrowths; Ti-bearing magnetite.
– Plagioclase, pyroxene, olivine, garnet, biotite, apatite, ulvospinel, rutile, pyrrhotite
(14.b) Magnetite, Ti-bearing magnetite, ilmenite as discrete grains and as exsolution intergrowth in magnetite in various proportions.
– Plagioclase (commonly labradorite), olivine, pyroxene, apatite; small amounts of sphene, rutile, spinel, pyrite, chalcopyrite, pyrrhotite

AGE, HOST ROCKS

(14.a) Mainly Proterozoic (mostly in the Grenville Province).
(14.b) All ages; mostly Precambrian.

AGE, ORE

In part contemporaneous with magmatic layering of host intrusions, in part syngenetic with late stage autointrusions.

GENETIC MODEL

Iron and titanium oxide phases separated by crystal settling, or filter pressing during crystallization of anorthositic (14.a) and gabbroic (14.b) magmas, forming syngenetic layers and segregations, as well as massive oxide autointrusions in lithified or partly solidified gabbroic anorthosite and genetically related host rocks. The separation of late stage, Fe-Ti-P-rich immiscible liquid has been proposed to account for the autointrusions.

ORE CONTROLS, GUIDES TO EXPLORATION

1. Large belts of anorthositic intrusions.
2. The host intrusive complexes commonly show a range of differentiated phases from anorthosite and norite to syenite.
3. Abundance of titanium-bearing magnetite and iron-titanium oxide minerals in stream sediments and derived sands has been used effectively in exploration.
3. Higher specific gravity of mafic intrusions in contrast to granitic gneisses.
4. Magnetic and gravity surveys are useful exploration methods.

(14.a) Ilmenite deposits tend to be associated with the less magnesian, andesine-type of anorthosite intrusion or phases thereof. Fe:Ti ratios of disseminated oxides less than 3, commonly about 2, are favourable (Fe:Ti ratios of disseminated oxide minerals tend to be similar to those in associated massive oxide ores). Massive ilmenite bodies commonly have "negative" magnetic anomalies.
(14.b) Titaniferous magnetite deposits tend to be associated with the magnesian, labradorite-type of anorthosite intrusion or phase thereof, and the Fe:Ti ratio tends to be greater than 3. Intrusions bearing significant Fe-Ti oxide concentrations are characterized by broad magnetic "highs".

AUTHORS

G.A. Gross, E.R. Rose (see Plate 12, page 11)

15. INTRUSION—ASSOCIATED GOLD

15.a Sub-alkalic Felsic
15.b Alkalic
15.c Mafic

COMMODITIES	Au (Ag)
EXAMPLES: **Canadian** – ***Foreign***	**(15.a)** Lamaque and Belmoral at Val d'Or, and Camflo and Barnat at Malartic, Que.; Wilmar at Red Lake, and Pamour Schumacher Division (formerly McIntyre) and Hollinger at Timmins, Ont.; – *Charters Towers and Meekathara, Australia* **(15.b)** Kirkland Lake camp, and Young-Davidson at Matachewan, Ont. **(15.c)** Howey-Hasaga at Red Lake, Ont.; San Antonio, Man.; – *Kalgoorlie Golden Mile deposits, Australia*
IMPORTANCE	This type accounts for approximately 35% of total Canadian gold production (excluding by-product gold) to date, the subtypes individually accounting for: **(15.a)** about 23% **(15.b)** about 11% **(15.c)** about 1%
TYPICAL GRADE, TONNAGE	Most of the significant mines of subtypes 15.a and 15.c produced 1 to 5 million tonnes grading 7 to 17 g/tonne. The largest Canadian producer of subtype 15.a is the Hollinger (more than 60 million tonnes, 10 g/tonne) and the largest Canadian producer of subtype 15.c is the San Antonio (4.1 million tonnes, 9 g/tonne). The Kirkland Lake camp (excluding Upper Canada mine) is of subtype 15.b and has produced at least 46 million tonnes grading 15.4 g/tonne. Au/Ag ratios ranging from 4 to 9 for deposits in this camp are lower than those of most other gold deposit types.
GEOLOGICAL SETTING	Archean greenstone belts; many deposits are associated with major structural breaks. Deposits are located in or near plutonic bodies: in general only this spatial association and not an intimate genetic relationship is evident.
HOST ROCKS OR MINERALIZED ROCKS	**(15.a)** Tonalite-granodiorite-quartz monzonite stocks, plugs and dykes and surrounding country rocks. Some (e.g. McIntyre) probably are subvolcanic intrusions. **(15.b)** Syenitic intrusions and related (?) alkalic volcanic rocks, and metavolcanic and metasedimentary country rocks. **(15.c)** Subvolcanic diorite to gabbro bodies. Deposits in Canada that contain both disseminated and fracture-controlled mineralization and might be considered to constitute a "porphyry gold" class (Franklin and Thorpe, 1982) are hosted by plutons belonging to subtypes 15.a and 15.b (e.g. Young-Davidson, Ont. and No. 6 Zone, Barnat Mine, Que.). The "porphyry gold" type of deposit is closely related in characteristics and probably in genesis to deposits of the porphyry copper type 17.
ASSOCIATED ROCKS	**(15.a)** Rocks characteristic of any part of an Archean greenstone belt. **(15.b)** Alkalic volcanic rocks and locally derived sedimentary rocks; some ultramafic volcanic and intrusive rocks. **(15.c)** Mafic volcanic rocks and greywacke.
FORM OF DEPOSIT, DISTRIBUTION OF ORE MINERALS	Deposit form is commonly controlled by individual faults or shear zones (Kirkland Lake, Camflo, McIntyre in part); en echelon vein systems (Lamaque, Howey-Hasaga); less commonly broadly disseminated, controlled by minor fractures (Young-Davidson).
MINERALS: Principal ore minerals – ***Associated minerals***	Native gold; at Kirkland Lake, Au-Ag tellurides as well. – *Quartz, carbonates, chlorite, sericite, pyrite, altaite, less commonly other sulphides and tellurides, tourmaline*
AGE, HOST ROCKS	Most are Archean, a few are Aphebian.
AGE, ORE	In general, probably not significantly younger than the last period of deformation and pluton emplacement. Not necessarily coeval with the spatially associated intrusions.
GENETIC MODEL	For the alkalic subtype (Kirkland Lake) and some deposits of the sub-alkaline felsic subtype (McIntyre, Wilmar) there may be a close genetic relationship to the associated intrusions, but the relationships are not understood in detail. For other deposits there is no well-established model; the intrusions may simply serve as structurally and chemically favourable traps, or as possible "heat engines" responsible for the establishment of circulating hydrothermal systems.
ORE CONTROLS, GUIDES TO EXPLORATION	Presence of small intrusions near major structural breaks is a broad guide (Val d'Or, Kirkland Lake). Fyfe and Kerrich (1979) have suggested that gold-bearing felsic intrusions as a whole exhibit anomalous enrichment of Na_2O, total volatiles, and $\delta^{18}O$, even in regions remote from individual gold-bearing veins, compared to "normal" values for these parameters in unmineralized felsic bodies. In some cases, large pyritic alteration zones are present.
AUTHORS	R.I. Thorpe, J.M. Franklin

16. CARBONATITE—HOSTED DEPOSITS

16.a Nephelinitic Carbonatite
16.b Ultramafic Carbonatite

COMMODITIES

(16.a) Nb, REE (P, U)
(16.b) Cu, vermiculite (Fe, P, U, Zr)

EXAMPLES: Canadian – *Foreign*

(16.a) St. Honoré (Nb) and Oka (Nb), Que.; Nemegosenda Lake (Nb), Lackner Lake (Nb), Martison Lake (P, Nb), and Cargill Township (P), Ont. – *Mountain Pass (REE), California; Araxa (Nb), Brazil; Fen (Nb), Norway*
(16.b) – *Palabora (Cu, Fe, P, Zr, U, vermiculite), South Africa*

IMPORTANCE

Canada: this deposit type is virtually the only source of niobium; accounted for about 17% of world production (1978). Potential source of phosphate.
World: **(16.a)** Accounts for nearly all production and reserves of niobium, and a small proportion of phosphate production. Mountain Pass accounts for most of the non-communist world's REE production.
(16.b) Palabora is an unusual but large producer of copper, vermiculite, zirconium, and apatite, as well as some iron and uranium.

TYPICAL GRADE, TONNAGE

Canada: Oka – 25.6 million tonnes, 0.44% Nb_2O_5 (proven reserves, 1974)
St. Honoré – 38.5 million tonnes, 0.70% Nb_2O_5 (proven plus estimated reserves, 1977)
Cargill – 62.5 million tonnes, 19.6% P_2O_5
Martison Lake – 140 million tonnes 20% P_2O_5, 0.35% Nb_2O_5
Foreign: Some of the largest known deposits include
(16.a) Araxa – 300 million tonnes, 3% Nb_2O_5
Mountain Pass – 5 million tonnes of contained rare earth oxides
(16.b) Palabora – initial reserves were estimated to be greater than 300 million tonnes of about 0.7% Cu

GEOLOGICAL SETTING

Deposits occur in circular to elliptical (rarely lens-like or vein-like in deformed rocks) carbonatite complexes, which tend to cluster near major normal or wrench faults. Intense assimilation and metasomatism is typical, to the point where distinction between intrusion and country rock is obscure.
(16.a) The nephelinitic carbonatite type of complex has a prominent fenitized (alkali metasomatized) aureole.
(16.b) The ultramafic carbonatite type tends to have more limited associated fenitization.

HOST ROCKS OR MINERALIZED ROCKS

Mineralization tends to be associated with one or two phases of the carbonatite complex, but differs from one complex to another. Known hosts include dolomitic carbonatite, calcitic carbonatite, and syenitic and pyroxenitic rocks. At Palabora, foskorite (a magnetite-olivine-apatite-phlogopite rock) is an important host rock.

ASSOCIATED ROCKS

Other phases of the carbonatite complexes, e.g. peridotite, pyroxenite, alnöite, urtite, ijolite, nepheline syenite, diorite.

FORM OF DEPOSIT, DISTRIBUTION OF ORE MINERALS

Lenses and layers, commonly crescent-shaped, distributed in an annular manner, reflecting the concentric pattern of lithologic units in carbonatite complexes. Ore bodies tend to be restricted to one or a few units, but weaker mineralization may extend to others. Ore minerals tend to be more or less uniformly disseminated. In some cases, mineralization extends into the fenite aureole. At Palabora, copper occurs in both layered concentric carbonatite and massive transgressive carbonatite.

Residual apatite concentrations form irregular blanket-like layers draped over some eroded and weathered carbonatite intrusions. At Araxa, the niobium ore consists of a residual pyrochlore concentration.

MINERALS: Principal ore minerals – *Associated minerals*

(16.a) Pyrochlore-microlite, columbite-tantalite, bastnaesite, apatite.
(16.b) Chalcopyrite, bornite, cubanite, vermiculite, magnetite, apatite, uranothorianite, baddeleyite.
– *Calcite, dolomite, hematite, magnetite and a variety of the rock-forming, accessory, and minor ore minerals found in carbonatites and related alkaline rocks*

AGE, HOST ROCKS

Various ages. Commonly postorogenic (Oka, about 0.12 Ga; St. Honoré, 0.64 Ga; Nemegosenda, 1.01 Ga; Lackner Lake, 1.09 Ga; Cargill, about 1.8 Ga; Palabora, > 2.06 Ga; Mountain Pass, Precambrian).

AGE, ORE

Syngenetic with host rocks. Copper at Palabora may in part be slightly younger. Residual deposits are related to younger unconformities.

GENETIC MODEL

Ores are primary constituents of carbonatite intrusive complexes, emplaced under generally tensional conditions, apparently localized crudely along crustal fracture systems. These systems may occur at considerable distance from the related orogenic activity. The intrusions are characterized by low silica and high CO_2/H_2O ratio, and are probably generated at appreciable depth in the mantle, though perhaps profoundly modified in the crust.

Subsequent weathering may, under favorable conditions, produce residual apatite and pyrochlore concentrations, and conversion of phlogopite to vermiculite.

ORE CONTROLS, GUIDES TO EXPLORATION

1. Typically, deep crustal fracture systems appear to control carbonatite complex distribution.
2. Particular phases of carbonatite complexes contain the potential ore zones.
3. Blanket-like residual deposits resulting from weathering of the carbonatites may contain apatite and pyrochlore concentrations of economic interest.
4. Annular magnetic and radiometric anomalies are particularly useful guides to unidentified carbonatite complexes.

AUTHORS

K.R. Dawson, K.L. Currie

Figure 18. *17.a Porphyry Copper, Molybdenum, Tungsten. Granisle mine, British Columbia. Intermineral porphyry dyke cuts fine grained porphyry, magnetite- and chalcopyrite-bearing quartz veins, and is cut by chalcopyrite-bearing quartz veins. Sample is 25 cm long. Sample: R.V. Kirkham (GSC 201531-M).*

Figure 19. *17.b Porphyry Copper, Molybdenum, Tungsten. Pit No. 2, Copper Mountain, British Columbia. Typical erratic, discontinuous veins, lenses, and disseminations of chalcopyrite (white) in intensely altered monzonite. Sample is 17 cm high. Sample: R.V. Kirkham (GSC 203886-G).*

17. PORPHYRY COPPER, MOLYBDENUM, TUNGSTEN

17.a Calc-Alkalic – associated Copper, Molybdenum

17.b Alkalic – associated Copper

17.c Calc-Alkalic – associated Molybdenum, Tungsten

COMMODITIES

(17.a) Cu, Mo (Au, Ag, Re)
(17.b) Cu (Au, Ag)
(17.c) Mo, W, (Sn, Bi)

EXAMPLES: Canadian – *Foreign*

(17.a) Island Copper, Highland Valley, Brenda and Granisle, B.C. – *Butte, Montana; Bingham, Utah; Morenci, Arizona; Cerro Colorado, Panama; Chuquicamata, El Teniente and El Salvador, Chile; Panguna, New Guinea; Atlas, Philippines; Sar Cheshmeh, Iran; Kounrad, U.S.S.R.*
(17.b) Copper Mountain, Afton and Galore Creek, B.C.
(17.c) Endako, Alice Arm and Adanac, B.C.; Logtung, Yukon; Mount Pleasant, N.B. – *Climax and Henderson, Colorado; Yanchuling and Xinglokeng, China*

IMPORTANCE

Canada: accounts for 30 to 40 per cent of copper production and all molybdenum production. Gold and silver are important byproducts. About 60 per cent of copper reserves and 99 per cent of molybdenum reserves.
World: Approximately 60 per cent of world's copper reserves and 99 per cent of world's molybdenum reserves.

TYPICAL GRADE, TONNAGE

(17.a) 0.2 to 1% Cu, 0.01 to 0.05% Mo; 50 to 1000 million tonnes (larger deposits are known outside Canada).
(17.b) 0.5 to 1.0% Cu, 0.2 to 1.0 g Au/tonne; 30 to 130 million tonnes.
(17.c) 0.08 to 0.20% Mo, nil to 0.20% W; 50 to 500 million tonnes.

GEOLOGICAL SETTING

Associated with epizonal or mesozonal, felsic to intermediate, silica-saturated and -undersaturated intrusions:
(17.a) near consumptive plate margins in both island and continental arc settings.
(17.b) near consumptive plate margins in intra-arc rift systems.
(17.c) in elevated continental plateaus and rifts, arc systems (?) and possibly over isolated mantle plumes.

Most Precambrian porphyry deposits occur within, or near the margins of, greenstone belts.

HOST ROCKS OR MINERALIZED ROCKS AND ASSOCIATED ROCKS

Highly variable. Mineralized rocks include genetically related intrusions and related and/or unrelated surrounding country rocks that include a variety of sedimentary, volcanic, intrusive and metamorphic rocks. Genetically-related intrusions are characteristically epizonal to mesozonal porphyritic rocks of the following compositional suites:
(17.a) diorite to granite, mainly granodiorite and quartz monzonite; extrusive equivalents and pyroclastic andesite breccias are common country rocks.
(17.b) diorite to nepheline syenite, mainly monzonite and syenite; related alkalic volcanic rocks are common country rocks.
(17.c) leucocratic quartz monzonite and granite, including sediment-derived anatectic granite ("S-type").

Hydrothermal alteration is extensive and consists typically of an inner potassic zone (e.g., biotite, K-feldspar) closely associated with economic sulphides, surrounded by propylitic alteration (e.g., epidote, chlorite, calcite) associated with pyrite. Phyllic (e.g., quartz, sericite, pyrite) and argillic (e.g., quartz, kaolin, montmorillonite) alteration can be either part of the zonal pattern between the potassic and propylitic zones or can be somewhat irregular or tabular younger zones superimposed on older alteration and sulphide assemblages.

FORM OF DEPOSIT, DISTRIBUTION OF ORE MINERALS

Variable forms: irregular, oval, solid or hollow cylindrical, inverted cup-shaped (in places, stacked). Large (typical dimensions 1000s x 1000s x 100s of metres). Mineralization dominantly structurally-controlled; mainly stockworks, veins, vein sets, breccias, disseminations and replacements. Sulphides and related alteration are in many instances zoned in crudely concentric patterns about mineralizing "centres". However, because mineralization and alteration zones of different ages are offset spatially in some deposits, overall patterns can be complex and irregular.

MINERALS: Principal ore minerals – *Associated minerals*

(17.a) Chalcopyrite, bornite, chalcocite, enargite, other copper minerals, molybdenite.
– *Pyrite, other sulphides, magnetite, quartz, biotite, K-feldspar, anhydrite, muscovite, clay minerals, epidote, chlorite*
(17.b) Chalcopyrite, bornite, other copper minerals, native gold.
– *Pyrite, other sulphides, magnetite, quartz, biotite, K-feldspar, anhydrite, epidote, chlorite, scapolite, albite, calcite, garnet*
(17.c) Molybdenite, scheelite, wolframite, cassiterite.
– *Pyrite, bismuthinite, other sulphides, native bismuth, magnetite, quartz, K-feldspar, biotite, muscovite, clay minerals, fluorite, topaz*

AGE, HOST ROCKS

Archean to Recent

AGE, ORE

Archean to Recent, but most economic deposits are Mesozoic or Tertiary. Penecontemporaneous with related intrusions.

GENETIC MODEL

Magmatic-hydrothermal; large volumes of highly saline aqueous fluids under high pressure, migrating upwards and outwards from a magmatic centre. Extensive fracturing (± brecciation) associated with ore formation. Mineralization and alteration, along with associated intrusive activity, occur in multiple stages, commonly superimposed or overlapping. Early stages of ore formation dominated by magmatic water; during waning stages, mineralization and alteration dominated by influx of meteoric water. Porphyry copper deposits may be enriched by subsequent weathering but generally this is not an important feature in Canadian deposits.

ORE CONTROLS, GUIDES TO EXPLORATION

1. Epizonal to mesozonal, felsic to intermediate, typically porphyritic intrusions, that have extensive related hydrothermal features such as sulphides, alteration, veins, stockworks and breccias.
2. Alteration related to hydrothermal activity is commonly zoned.
3. Metal and mineral zonation; tends toward concentric patterns, but may be complex and somewhat irregular.
4. Extensively developed favourable structures such as stockworks, fractures, fault systems, and breccias.
5. An enriched zone developed by weathering of the primary sulphide deposit can have much higher copper grade, thereby enhancing the possibility of economic exploitation.
6. Surficial and rock geochemical studies may be effective exploration guides to ore.

AUTHORS

R.V. Kirkham, W.D. Sinclair

Figure 20. *17.c Porphyry Copper, Molybdenum, Tungsten. Kitsault mine, British Columbia. Typical intermineral porphyry dyke cutting highly altered granodiorite porphyry. The dyke truncates some molybdenite-bearing quartz veins and is cut by others. In the upper left corner of the specimen is a late quartz-carbonate-sphalerite-galena vein. Sample: R.V. Kirkham (GSC 202871-F).*

Figure 21. *17.c Porphyry Copper, Molybdenum, Tungsten. Kitsault mine, British Columbia. Typical molybdenite-bearing quartz vein stockwork cutting porphyritic granodiorite and aplite dykelet. Photo: R.V. Kirkham (GSC 3-16-72).*

18. VEIN—STOCKWORK TIN

COMMODITIES	Sn (W, Mo, Cu, Zn, Ag, Bi)
EXAMPLES: **Canadian** – ***Foreign***	East Kemptville, N.S.; Mount Pleasant (Sn zones), N.B.; – *Cornwall, England; Altenberg, East Germany; Hub, Czechoslovakia; Catavi, Oruro and Chorolque, Bolivia; Aberfoyle and Taronga, Australia; Akenobe, Japan*
IMPORTANCE	Canada: no production at present; potential production from East Kemptville, N.S. World: approximately 50% of production of tin in 1973.
TYPICAL GRADE, TONNAGE	1 to 5 million tonnes at 0.85 to 1.4% Sn (e.g. Mount Pleasant Sn zones, N.B.; Aberfoyle, Australia; Geever Mine, Cornwall, England). 15 to 80 million tonnes at 0.2 to 0.3% Sn (e.g. Altenberg, East Germany; East Kemptville, N.S.; Taronga, Australia; Redmoor and Hemerdon, Cornwall, England). 100 to 1000 million tonnes at 0.2 to 0.5(?)% Sn (e.g. Catavi and Oruro, Bolivia).
GEOLOGICAL SETTING	Cupolas or domes associated with hypabyssal granitic intrusions, shallow to subvolcanic intrusions and related breccia pipes; in orogenic belts, particularly in miogeosynclinal areas.
HOST ROCKS OR MINERALIZED ROCKS	Peraluminous granitic rocks ("S-type"); silica- and potash-rich calc-alkaline granitic rocks and aphanitic counterparts. Mineralization may extend into surrounding country rocks.
ASSOCIATED ROCKS	Miogeosynclinal sedimentary rocks and minor volcanic rocks, older granitic and metamorphic rocks.
FORM OF DEPOSIT, DISTRIBUTION OF ORE MINERALS	Tabular to irregular veins and vein systems; oval, cylindrical and irregular stockworks. Mineralization occurs in veins and veinlets, along fractures and disseminated in altered wall rocks.
MINERALS: Principal ore minerals – ***Associated minerals***	Cassiterite, stannite, other tin sulphides and sulphosalts, wolframite, molybdenite, bismuthinite, chalcopyrite, sphalerite. – *Pyrite, pyrrhotite, hematite, arsenopyrite, tourmaline, topaz, fluorite, muscovite, beryl, lepidolite, zinnwaldite, biotite, chlorite, quartz, K-feldspar, clay minerals*
AGE, HOST ROCKS	Precambrian to Tertiary.
AGE, ORE	Penecontemporaneous with related intrusions. Precambrian to Tertiary; ages clustered mainly around Silurian-Devonian, Permian, Jurassic-Cretaceous, and Tertiary.
GENETIC MODEL	Tin was concentrated in B- and/or F-rich fluid or vapour phases related to silica- and potash-rich granitic magmas that were probably derived by anatectic melting of sedimentary rocks ("S-type" granites). Within these intrusions and their adjacent country rocks tin was deposited (commonly in multiple stages) in faults, fractures, breccias, and altered wall rocks.
ORE CONTROLS, GUIDES TO EXPLORATION	1. Commonly associated with granitic rocks anomalous in Sn, F, Cl, Li, B, Rb; look for the intrusive phases enriched in these elements, particularly local cupolas, domes and subvolcanic intrusions. 2. Mineralization is typically fracture-controlled; look for high fracture density, also crosscutting mineralized fractures or vein systems that indicate multiple mineralization stages. 3. Tin mineralization is commonly associated with quartz-muscovite-topaz ± fluorite alteration of feldspars and biotite (greisenization). 4. Mineralogical and metal zonation may be developed on a small scale (single veins or vein systems) and/or a larger scale (districts).
AUTHORS	W.D. Sinclair, R.V. Kirkham

Figure 22. *18 Vein-Stockwork Tin. East Kemptville, Nova Scotia. Greisenized zone of quartz, topaz and muscovite forms an alteration envelope approximately 2 cm wide around a quartz veinlet cutting sericitized granite, East Kemptville tin deposit. Such zones contain cassiterite, as well as pyrite, sphalerite, chalcopyrite, pyrrhotite, minor wolframite and other sulphide minerals. Photo: W.D. Sinclair (GSC 203886-Q).*

Figure 23. *18 Vein-Stockwork Tin. Todd Mountain, New Brunswick. Subparallel swarm of cassiterite-bearing quartz veins, veinlets and fractures cutting foliated biotite hornfels. Photo: W.D. Sinclair. (GSC 203425-I)*

19. SKARN DEPOSITS

19.1 SKARN TUNGSTEN

COMMODITIES
W (Cu, Zn, Mo, Bi)

EXAMPLES: Canadian – *Foreign*
Cantung, N.W.T.; Mactung, Yukon; Salmo District, B.C. – *Bishop District, California; Sangdong, South Korea; King Island, Tasmania, Australia*

IMPORTANCE
Canada: almost 100% of tungsten produced is from skarns. Represents about 5% of annual world production.
World: skarns are the most common type of economic tungsten deposits, and account for an estimated 30% of total production.

TYPICAL GRADE, TONNAGE
Size is highly variable; grade generally ranges from 0.4 to 2.0% WO_3. Five of the largest deposits are:

Mactung - 57 million tonnes, 0.95% WO_3
Cantung - 9 million tonnes, 1.4% WO_3, 0.2% Cu
King Island - 13 million tonnes, 0.8% WO_3
Sangdong - 3.8 million tonnes, 1.75% WO_3
Bishop District - ca. 10 million tonnes, 0.5% WO_3

GEOLOGICAL SETTING
In widespread thermal aureoles at contacts between felsic intrusive and calcareous sedimentary rocks. Cordilleran skarns are localized where upper Mesozoic plutons discordantly intrude Paleozoic outer shelf carbonate-pelite sequences. Tungsten skarns are emplaced in a generally deeper, higher temperature and more reduced environment than copper and zinc-rich skarns, as deduced from extensive thermal aureoles, coarse grained intrusive texture, presence of migmatitites, low ferric/ferrous ratios, and abundant carbon and pyrite in host rocks.

HOST ROCKS OR MINERALIZED ROCKS
Contact metamorphosed equivalents of relatively pure limestone beds, impure limestones and calcareous to carbonaceous pelites: skarn and calc-silicate and biotite-pyrite hornfels. Cordilleran skarns develop preferentially in the lowest thick limestone bed of a Proterozoic to Paleozoic stratigraphic sequence. Cambrian limestone, underlain by, and interbedded with, pelite and carbonaceous shale, is a typical setting.

ASSOCIATED ROCKS
Calc-alkalic felsic stocks, plutons or batholiths: quartz monzonite most common, quartz diorite least common. Plutons are commonly coarse grained, porphyritic and unaltered, but border phases may be argillized, greisened, or tourmalinized locally. Stockwork quartz veining is not extensive, but more abundant within intrusive rock than in skarn. Few associated porphyritic dykes, but aplite and pegmatite dykes common. Breccia pipes, intrusive and shatter breccias absent.

FORM OF DEPOSIT, DISTRIBUTION OF ORE MINERALS
Commonly as essentially stratiform units tens to hundreds of metres away from an intrusive contact; also as semiconcordant to discordant bodies immediately adjacent to an intrusive contact; less commonly as xenoliths, pendants and screens within a plutonic phase. Scheelite occurs in the stratiform almandine-hedenbergite skarn assemblage which overprints the contact calc-silicate hornfels, but it may also be erratically redistributed in association with retrograde amphibole-biotite alteration zones.

MINERALS: Principal ore minerals – *Associated minerals*
Scheelite.
– Chalcopyrite, sphalerite, molybdenite, pyrrhotite, pyrite (late); common gangue minerals are pyroxene (diopside-hedenbergite) and garnet (grossular-andradite-almandine), calcite, dolomite, quartz, biotite; less abundant minerals include vesuvianite, sphene, wollastonite, anorthite, fluorite, and common silicate alteration minerals

AGE, HOST ROCKS
Late Proterozoic to Triassic, predominantly Cambrian.

AGE, ORE
Penecontemporaneous with associated intrusive rocks: Late Jurassic to Early Tertiary in Cordillera.

GENETIC MODEL
Tungsten and associated metals and sulphur may have been derived from both pluton and host pelite, by a magmatic-hydrothermal fluid, by convecting groundwater or formational water or by a combination of both. Scheelite deposition was controlled mainly by prograde metasomatic reactions of ore fluid with calcium carbonate host in the thermal aureole of the pluton, and accompanied main stage almandine-hedenbergite skarn development, superimposed on the calc-silicate hornfels aureole. Contemporaneously, a wollastonite-vesuvianite assemblage developed in marble, and a pyroxene-epidote-plagioclase assemblage in plutons and pelitic hornfels. Subsequent cooling and influx of meteoric water caused hydrous retrograde alteration of calc-silicate minerals, redistribution of scheelite (including depletion and upgrading), and deposition of sulphides.

ORE CONTROLS, GUIDES TO EXPLORATION
1. Extensive hornfels zone adjacent to an exposed pluton, or overlying a buried one.
2. Relatively thick, pure and impure limestone beds.
3. Shallowly dipping pluton/limestone contacts.
4. Structural and stratigraphic traps in pelite-carbonate host.
5. Irregularities in pluton/limestone contact, particularly reentrants and troughs.
6. Stockwork fracturing along pluton/limestone contact.

AUTHOR
K.M. Dawson

19.2 SKARN ZINC-LEAD-SILVER

COMMODITIES	Zn, Pb, Ag (Cu, W)
EXAMPLES: Canadian – *Foreign*	Cassiar and Mount Billings Batholiths, Yukon; HPH and Zip deposits, B.C.; Meat Cove, N.S. – *Central Mining District (Hanover), New Mexico; Santa Eulalia, Chihuahua, Mexico; Trepca, Yugoslavia; Yeonhua, Korea*
IMPORTANCE	Canada: numerous small subeconomic deposits in northern Cordillera, Vancouver Island, and Appalachians. World: signficant past and current production from Mexico, New Mexico, Yugoslavia, Korea, Japan, California, Argentina, U.S.S.R., China.
TYPICAL GRADE, TONNAGE	Canada: less than 1 million tonnes, 10 to 20% Zn or Zn+Pb, 30 to 60 g Ag/tonne. Foreign: some examples of large deposits: Santa Eulalia - 29 million tonnes, 11% Zn, 10% Pb, 200 g Ag/tonne Central Mining District - 18 million tonnes, 14% Zn, 0.3 to 4% Pb, 1% Cu, 70 to 140 g Ag/tonne Stri Trg Mine, Trepca - 12.5 million tonnes, 3.8% Zn, 8.6% Pb, 0.2% Cu, 140 g Ag/tonne Yeonhua I and II - 9.6 million tonnes, 6.6% Zn, 3.0% Pb, 0.1% Cu
GEOLOGICAL SETTING	Either in thermal aureoles at contacts between felsic to intermediate intrusive and calcareous sedimentary rocks or along structural pathways in unmetamorphosed rocks distant from an intrusive source. The thermal aureoles are less extensive than in tungsten- and copper-rich skarns. Cordilleran skarns are localized typically where upper Mesozoic plutons discordantly intrude lower Paleozoic outer shelf carbonate-pelite sequences, except on Vancouver Island where the intruded rocks are Paleozoic and Mesozoic oceanic arc-type volcanic-carbonate sequences. In Nova Scotia, the intruded rocks are Precambrian shelf carbonate sequences.
HOST ROCKS OR MINERALIZED ROCKS	Most Cordilleran skarns are in contact metamorphosed equivalents of relatively pure limestone beds, impure limestones and calcareous pelites, or regionally metamorphosed equivalents: skarn and hornfels (calc-silicate and biotite-pyrite). On Vancouver Island, skarns are in thick limestone beds between basaltic lava flows and pyroclastic rocks. In Nova Scotia skarns occur in intensely metamorphosed Precambrian marble.
ASSOCIATED ROCKS	Felsic to intermediate stocks or plutons: quartz monzonite most common, quartz diorite less common; also leucogranitic plutons and minor intrusions (syenite at Meat Cove). Dykes and sills locally abundant. Border phases may be argillized, and locally greisened or tourmalinized. Quartz veining more abundant within intrusive rocks than in skarn. Intrusive rock is generally not in contact with skarn.
FORM OF DEPOSIT, DISTRIBUTION OF ORE MINERALS	Form variable: broadly stratiform skarns follow limestone bedding near plutonic contacts; semiconcordant to elongate discordant bodies occur commonly at lithologic and structural contacts at some distance from plutonic and dyke margins; thin conformable skarn layers in biotite schist. Proximal zinc-lead skarns tend to have relatively high copper and tungsten contents; distal skarns tend to be rich in manganese, silver and lead.
MINERALS: Principal ore minerals – *Associated minerals*	Sphalerite (mainly Fe-rich), more abundant than galena. – *Chalcopyrite, scheelite, pyrrhotite, pyrite, arsenopyrite. Fe- and Mn-rich calc-silicate minerals: manganoan hedenbergitic pyroxene predominates; andraditic garnet, diopside, hastingsite, epidote, magnetite, vesuvianite, wollastonite. Retrograde minerals: manganiferous actinolite, chlorite, epidote, ilvaite, rhodonite, fluorite, calcite and quartz. At Meat Cove, assemblage of syenitic affinity includes silica undersaturated minerals, e.g. scapolite*
AGE, HOST ROCKS	Late Proterozoic to Cretaceous.
AGE, ORE	Same as associated intrusive rock: Jurassic to Early Tertiary in Cordillera. Conformable skarns at Mount Billings may be related to an older regional metamorphic event.
GENETIC MODEL	Zn, Pb and associated metals may be derived from both pluton and country rocks by a magmatic-hydrothermal system, by convecting groundwater or formational water, or by a combination of both. Metal transport may be effected by a broad range of chemical mechanisms, including fluoride and chloride complexing. Metal deposition controlled mainly by reaction of ore fluid with calcium carbonate in country rocks. Zonation of metals into early, proximal, iron-deficient, tungsten-copper-rich assemblages, intermediate Pb-Zn-Ag-rich assemblages, and late, distal iron-manganese-rich assemblages probably reflects evolution of metasomatic fluid composition away from an igneous heat source. Conformable base metal skarns may develop penecontemporaneously with metamorphic formation of calcareous schist, adjacent to synmetamorphic granitic bodies in high grade metamorphic-migmatitic terrane; they do not represent pre-metamorphic stratiform sulphide deposits.
ORE CONTROLS, GUIDES TO EXPLORATION	1. Relatively thick limestone beds. 2. Shallow-dipping pluton-limestone contacts. 3. Structural and stratigraphic traps in host rocks. 4. Irregularities in pluton-limestone contact, particularly reentrants and troughs. 5. Stockwork fracturing along pluton/limestone contact. 6. Limestone - leucogranite contacts in high grade metamorphic-migmatitic terrane. 7. Some zinc skarns are controlled by fault and/or dyke contacts at some distance from the intrusive contact.
AUTHORS	K.M. Dawson, D.F. Sangster

19.3 SKARN IRON

19.3.a Intrusion-associated (contact metasomatic)
19.3.b Stratiform in metamorphic terrane

COMMODITIES	**(19.3.a)** Fe(Cu, W, Sn) **(19.3.b)** Fe
EXAMPLES: Canadian – *Foreign*	**(19.3.a)** Tasu, Texada Island, B.C. – *Cornwall, Pennsylvania; Iron Springs, Utah; Tayeh, China; Magnitnaya, U.S.S.R.* **(19.3.b)** Marmora, Ont.
IMPORTANCE	Canada: small present and past producers; of minor significance. World: major producers in U.S.S.R., Peru, China.
TYPICAL GRADE, TONNAGE	Canada: deposits up to 30 million tonnes, averaging 40 to 50% Fe. Foreign: deposits up to 1000 million tonnes, U.S.S.R.; 100 to 200 million tonnes, China, Peru, U.S.A.
GEOLOGICAL SETTING	**(19.3.a)** Deposits commonly occur in mobile belts in metavolcanic-metasedimentary sequences that include limestone, at or near contacts with intrusions of varied compositions. **(19.3.b)** Deposits occur in highly metamorphosed and deformed sedimentary-volcanic sequences; intrusions may or may not be present.
HOST ROCKS OR MINERALIZED ROCKS AND ASSOCIATED ROCKS	Limestone, dolostone, skarn rock characterized by calc-silicate-bearing assemblages and various metavolcanic and metasedimentary rocks may all be mineralized. **(19.3.a)** Intrusive rocks ranging from quartz monzonite and syenite to gabbro are present, usually close to the ores, and in some cases are also mineralized. **(19.3.b)** Many deposits contain layered silicate-carbonate-quartz-magnetite (metasediments, metavolcanics) associated with more typical irregular masses of remobilized magnetite.
FORM OF DEPOSIT, DISTRIBUTION OF ORE MINERALS	Vary in form from irregular massive or disseminated patchy veins or dyke-like masses (mainly subtype 19.3.a), to more uniform tabular massive or layered conformable bodies (mainly subtype 19.3.b). Mineral distribution patchy to uniform.
MINERALS: Principal ore minerals – *Associated minerals*	Magnetite, hematite, martite, chalcopyrite. – *Pyrite, pyrrhotite; apatite, calc-silicate skarn assemblages, calcite, dolomite*
AGE, HOST ROCKS	Precambrian to Recent. In coastal areas of B.C. many deposits are associated with Late Triassic limestones and Jurassic intrusions.
AGE, ORE	**(19.3.a)** Penecontemporaneous with intrusions. **(19.3.b)** Iron concentration may be syngenetic sedimentary and/or metasomatic associated with regional metamorphism.
GENETIC MODEL	**(19.3.a)** Hydrothermal metasomatic replacement of wall rocks contemporaneous with emplacement of intrusions. Limestone localizes precipitation of magnetite probably by reacting to increase pH of ore fluids. **(19.3.b)** In some cases these deposits probably represent originally iron-rich sediments or mafic magmatic rocks, together with limestone, subsequently metamorphosed. In other cases, the present iron concentration may have been mainly attained by mobilization of iron and metasomatic replacement of favourable host rocks (typically calcareous) during metamorphism.
ORE CONTROLS, GUIDES TO EXPLORATION	**(19.3.a)** 1. The contact zones of intrusions. 2. Volcanic/sedimentary wall rocks that include limestone. **(19.3.b)** Highly metamorphosed and deformed supracrustal sequences that contain iron-rich and calcareous strata.
AUTHOR	G.A. Gross

19.4 SKARN COPPER

COMMODITIES	Cu (Fe, Au, Ag, Mo)
EXAMPLES: **Canadian – *Foreign***	Craigmont, Phoenix, Ingerbelle and Coast Copper, B.C.; Whitehorse Copper, Yukon; Gaspé Copper and Madeleine Mines, Que. – *Carr Fork, Utah; Ertsberg, Indonesia*
IMPORTANCE	Canada: skarn deposits account for approximately 10% of Canada's copper production and approximately 6% of its reserves. World: significant skarn copper deposits occur scattered around the world; in China, skarn is an important source of copper.
TYPICAL GRADE, TONNAGE	Canada: deposits mined range from 100 000 tonnes averaging 1.5% Cu to 200 million tonnes averaging 0.4%, but most of the economic deposits contain 1 to 20 million tonnes and average 1 to 2% Cu.
GEOLOGICAL SETTING	Most skarn copper deposits occur in mobile belts, in or near limestones or impure limestones, at or near contacts with mafic to felsic intrusions. The very large skarn copper deposits occur mainly in porphyry copper districts.
HOST ROCKS OR MINERALIZED ROCKS	Crystalline limestone, impure limestone, dolostone, Ca-Mg skarns and metavolcanic and meta-intrusive rocks.
ASSOCIATED ROCKS	The intrusions may be any of the following: gabbro to granite, diorite to syenite, or their porphyritic or aphanitic counterparts. Wall rocks to the intrusions include a variety of sedimentary, volcanic, and metamorphic rocks.
FORM OF DEPOSIT, DISTRIBUTION OF ORE MINERALS	Irregular, tabular, vein-like to peneconcordant bodies with patchy, massive and disseminated ore minerals.
MINERALS: Principal ore minerals **– *Associated minerals***	Chalcopyrite, magnetite, bornite, molybdenite. – *Pyrite, pyrrhotite, other sulphides, hematite, calcite, dolomite, quartz, andradite-grossularite, diopside-hedenbergite, epidote, actinolite, tremolite, chlorite, wollastonite, serpentine, K-feldspar, talc, biotite*
AGE, HOST ROCKS	Precambrian to Recent, predominantly Phanerozoic.
AGE, ORE	Precambrian to Recent, predominantly Phanerozoic. Penecontemporaneous with related intrusions.
GENETIC MODEL	Hydrothermal replacement, with lesser open space filling; products of hydrothermal systems related to emplacement of intrusions. Metals may have been derived from the intrusions, or from surrounding country rocks.
ORE CONTROLS, GUIDES TO EXPLORATION	1. Thick limestone beds in otherwise carbonate-poor sequences. 2. Close proximity to a magmatic-hydrothermal centre such as a porphyry copper deposit. Large, well-mineralized copper skarn deposits are rarely more than a few hundred metres from their associated intrusions. 3. Presence of channelways for ore-forming fluids; fractures, faults, stockworks, breccias, and permeable stratigraphic units. Skarns related to porphyry copper deposits tend to have greater vein and fracture densities than non-porphyry related skarns. 4. Metal zoning; in some deposits copper is concentrated near the skarn-marble contact whereas lead, zinc and silver tend to occur farther out in marble or other rocks. In porphyry copper districts, copper skarns with molybdenum tend to occur deep in the porphyry copper system whereas copper skarns with zinc occur farther away.
AUTHORS	R.V. Kirkham, W.D. Sinclair

(see Plate 13, page 11)

20. NEPHELINE— AND CORUNDUM—BEARING ALKALINE GNEISSES

20.a "Nepheline Syenite"
20.b Corundum
20.c Molybdenum

COMMODITIES

(20.a) "Nepheline Syenite" (alumina, white mica, corundum, sodalite, cancrinite, lapis lazuli)
(20.b) Corundum (gahnite)
(20.c) Mo

EXAMPLES: Canadian – ***Foreign***

(20.a) Blue Mountain, Haliburton-Bancroft-Renfrew belt, and Bigwood, Ont.; Frenchman's Cap and Trident Mountain, B.C.; Lake Harbour, Baffin Island, N.W.T. – *Vishnevogorsk, Ilmenogorsk and Mongol-Tuva, U.S.S.R.; Kishingar, India; Kipong and Dufo, Ghana; Makaraingobe, Malagasy Republic*
(20.b) Craigmont and Burgess Mines, Ont. – *Tambani and Port Herald, Malawi*
(20.c) Mount Copeland, B.C. – *Darkainle, Somalia*

IMPORTANCE

(20.a) Canada: probably 100% of world production of "nepheline syenite"; major reserves. Reserves elsewhere undefined. Probably 100% of world production of sodalite; small reserves. Minor production of lapis lazuli and cancrinite; small reserves.
(20.b) Canada: previously world's foremost supplier of corundum. Reserves may equal or exceed those of all other countries. Minor production of gem corundum; very small reserves.
(20.c) Canadian and other reserves are unimportant relative to other types of molybdenum deposits.

TYPICAL GRADE, TONNAGE

(20.a) Blue Mountain: production about 700 000 tonnes/year; tens of millions of tonnes at about 24% Al_2O_3, 15% alkalies. Davis Hill, Bancroft: tens of millions of tonnes at about 25% Al_2O_3, 16% alkalies. Monteagle deposit, Bancroft: 5 to 10 million tonnes at about 29% Al_2O_3, 9% alkalies, plus 8% white mica, 2.5% corundum. Bigwood: current exploration; about 20% Al_2O_3, 14% alkalies. Princess Sodalite Mine, Bancroft: small production from undefined reserves at 1 to 2% sodalite.
(20.b) Craigmont Mine: 5 million tonnes at about 7% corundum. Many other smaller deposits in Haliburton-Bancroft-Renfrew belt.
(20.c) Mount Copeland Mine: previous production - 154 000 tonnes at 0.75% Mo.

GEOLOGICAL SETTING

Gneissic belts, consisting of varied proportions of metavolcanic and metasedimentary rocks, commonly including appreciable quantities of marble and related Ca-Mg silicate rocks. Evidence of multiple orogenies and intrusion by various kinds of igneous rocks is common.

HOST ROCKS OR MINERALIZED ROCKS

(20.a) Feldspar- and feldspathoid-rich gneisses. Precursors might have been alkaline igneous rocks, volcanic or metasedimentary rocks, including marble and Ca-Mg silicate rocks.
(20.b) Alkaline gneisses and associated massive syenitoid metasomatized zones.
(20.c) Aplitic to pegmatitic segregations and dykes within or adjacent to alkaline gneisses.

ASSOCIATED ROCKS

Orthogneiss, paragneiss, amphibolite, marble, Ca-Mg-silicate rocks; less commonly, schist and quartzite. Also, migmatites and metasomatites derived from these. Intrusive rocks, predominantly felsic, but ranging to mafic and even ultramafic.

FORM OF DEPOSIT, DISTRIBUTION OF ORE MINERALS

(20.a) Conformable layers and lenses. Stratigraphic relations commonly preserved despite pre- and post mineralization folding. Mineral assemblages may vary greatly along or across strike. Localized pegmatitic recrystallized zones. Sodalite and cancrinite form patchy replacements.
(20.b) Corundum is minor constituent of some alkaline gneisses. Greater concentrations in patchy medium grained to pegmatitic syenitoid salients within or enveloping Al-rich gneisses.
(20.c) Molybdenite is minor constituent of some alkaline gneisses. Greater concentrations in aplitic to pegmatitic segregations and dykes within or adjacent to the gneisses.

MINERALS: Principal ore minerals – ***Associated minerals***

(20.a) "Nepheline syenite": generally nepheline, plagioclase (albite to andesine), microcline-perthite, and minor amounts of corundum, muscovite, scapolite, lazurite, sodalite, cancrinite.
– *Biotite, amphiboles, pyroxenes, magnetite, zeolites, calcite, garnet, titanite*
(20.b) Corundum, gahnite.
– *Alkali feldspars, nepheline, scapolite, muscovite, biotite, zeolites, magnetite, sulphides*
(20.c) Molybdenite.
– *Alkali feldspars, nepheline, pyroxene, biotite, zeolites, titanite*

AGE, HOST ROCKS

Most deposits are components of highly deformed Precambrian sequences. Ontario and Quebec sequences are presumed to be Proterozoic. The Mount Copeland sequence may be Early Cambrian.

AGE, ORE

Difficult to establish; the most recent metamorphic-metasomatic reworkings can be determined, e.g. Ontario and Quebec, about 1 Ga; Mount Copeland, 45 Ma.

GENETIC MODEL

Most deposits have been formed and modified in the course of several tectonic, intrusive and metamorphic events. More than one of the following proposed geneses may be valid, having in common only the requisite favorable metamorphic conditions:
A. Alkaline igneous rocks whose magmas were produced by (1) partial melting of mantle rocks; (2) differentiation products from crustal magma chambers; (3) desilication of granitic magma through assimilation of limestone; (4) palingenetic-metasomatic interactions at depth in supracrustal piles that included calcareous members; (5) remobilization of pre-existing rocks of similar composition.

B. Various kinds of pre-existing rocks metasomatized by alkali- and alumina-rich fluids resulting from (1) differentiation from subjacent magma; (2) palingenesis of underlying supracrustal or basement rocks; (3) desilication of juvenile or palingenetic fluids on traversing and reacting with silica-deficient rocks; (4) reactions between soda-rich evaporites and associated sedimentary rocks.
C. Alumina-rich laterites.
D. Analcime-rich sediments.

ORE CONTROLS, GUIDES TO EXPLORATION

(20.a) More or less continuous belts of rocks paralleling tectonic grain. Original compositional layering of supracrustal sequence. Marble and Ca-Mg-silicate rocks. Mineral assemblages and textures indicative of high temperature and pressure metamorphism. Sodalite and cancrinite occur as late hydrothermal replacements, generally fracture-controlled. Lapis lazuli occurs in partially metasomatized marble.
Exploration guides: Corundum stands out from weathered surfaces; feldspars are level and whitened; nepheline, sodalite, cancrinite and scapolite form pits and show white to bluish patina. White to red 'hydronephelite', green 'gieseckite'. Lateritization.
(20.b) Corundum is concentrated in massive syenitoid reaction zones within and enveloping Al-rich gneisses adjacent to granitic intrusions and granitic migmatites.
Exploration guides: Dark mineral: (1) if amphibole or pyroxene, no corundum; (2) if biotite, corundum likely; (3) if magnetite, corundum generally present. Eluvial and alluvial corundum.
(20.c) Molybdenite is concentrated in aplitic to pegmatitic segregations within alkaline gneisses and in adjacent dykes and sills.

AUTHOR

L. Moyd, Curator Emeritus, National Museum of Natural Sciences

21. UNCONFORMITY—ASSOCIATED URANIUM

COMMODITIES

U(Ni, Co, As, Se, Ag, Au, Mo)

EXAMPLES:
Canadian – *Foreign*

Key Lake, Rabbit Lake, and Cluff Lake - "D" ore body, Sask. – *Jabiluka I and II, Ranger I-ore bodies 1 and 3, Northern Territory, Australia*

IMPORTANCE

Canada: accounts for approximately 35% of current uranium production, but about 50% of current reserves.
World: represents about 15% of current reserves.

TYPICAL GRADE, TONNAGE

Canada: individual ore bodies range in size from very small up to 5 million tonnes of ore (up to 45 000 tonnes contained U). Typical grades range from 0.3 to 3% U, though some exceed 5%.
Australia: maximum reported size is 200 000 tonnes contained U, but grade is generally lower than in Canadian deposits.

GEOLOGICAL SETTING

Relatively undeformed, intracratonic, Helikian sedimentary basins (mainly comprising mature quartz sandstone) resting unconformably on intensely deformed Archean and Aphebian basement rocks. Deposits are associated with the unconformity where it is intersected by faults. In northern Saskatchewan, paleoregolith is present at the unconformity below the Athabasca Group sandstone.

HOST ROCKS OR MINERALIZED ROCKS

Ores occur in clay, sericite, and chlorite masses at the unconformity and along intersecting faults, and in clay- altered basement rocks (quartz-feldspar gneiss, graphite-bearing feldspathic quartzite) and kaolinized cover rocks (grey or multicoloured sandstone and shale). Most Australian deposits are in chloritized basement rocks.

ASSOCIATED ROCKS

Basement: graphitic schist and gneiss, coarse grained granitoid rocks, calc-silicate metasedimentary rock. Cover rock: sandstone and shale.

FORM OF DEPOSIT, DISTRIBUTION OF ORE MINERALS

Flattened cigar-shaped, high grade ore bodies oriented and distributed along the unconformity and faults, especially localized at unconformity-fault intersections. The high grade bodies grade outward into low grade stratiform disseminations and fracture fillings that occur in both basement and cover rocks, generally less than 200 m from the unconformity. The Australian deposits mostly occur below the unconformity.

MINERALS: Principal ore minerals
– *Associated minerals*

Pitchblende, coffinite, minor uranium oxides.
– Ni and Co arsenides and sulphides, native selenium and various selenides, native gold and Au tellurides, galena, minor molybdenite, Cu and Fe sulphides; clay minerals (principally illite and kaolinite), chlorite, quartz, graphite, carbonate. Radioactive phyllosilicates are known.

AGE, HOST ROCKS

Most basement host rocks are Aphebian, some may be Archean. Helikian cover rocks.

AGE, ORE

Helikian (1.281 ± 0.11 Ga, "oldest" pitchblende at Rabbit Lake); later remobilization, probably at several periods.

GENETIC MODEL

An early model proposed ascending hydrothermal solutions precipitating ore at the unconformity. Subsequent modelling invokes various combinations of the following mechanisms:

1. possible preconcentration of uranium during deposition of Aphebian (basement) sedimentary rocks, and during their anatexis,
2. further concentration of uranium during (Helikian) development of lateritic regolith,
3. mobilization of uranium from basement rocks, lateritic regolith and cover rocks by heated, oxidized (groundwater ? diagenetic?) solutions, and precipitation upon encountering a reducing environment at the unconformity/fault locus,
4. additional cycles of oxidation/mobilization and reduction/precipitation leading to redistribution and/or further concentration of uranium.

ORE CONTROLS, GUIDES TO EXPLORATION

1. Mineralization is at or near an unconformity between Helikian sandstone and underlying Aphebian and/or Archean basement rocks.
2. Intersection of the unconformity with basement faults that have been reactivated after deposition of the cover rocks.
3. Associated with graphitic basement rocks and grey-black and/or multicoloured shale and sandstone cover rocks.
4. Associated with intense clay, chlorite, and sericite alteration of basement and cover rocks.
5. Basement rocks with a higher than average uranium content.

AUTHORS

L.P. Tremblay, V. Ruzicka

Figure 24. *22 Arsenide Vein Silver, Uranium. Terra mine, Northwest Territories. Nickeline-native silver dendrites in ore. Sample is 11 cm long. Sample: Ralph Thorpe (GSC 201788-J).*

22. ARSENIDE VEIN SILVER, URANIUM

COMMODITIES	Ag, U (As, Co, Ni, Cu, Bi)
EXAMPLES: Canadian – ***Foreign***	Cobalt District, Ont.; Great Bear Lake District, N.W.T.
IMPORTANCE	Canada: historically this type was a major source of silver (Cobalt District, 19 million kilograms produced) and a significant source as well of uranium and radium (Great Bear Lake, 5200 tonnes U produced), and cobalt (Cobalt District, 20 000 tonnes produced). Currently of renewed interest for silver and cobalt. In recent years has accounted for about 8% of annual Canadian silver production, and less than 1% of cobalt production.
TYPICAL GRADE, TONNAGE	In the Cobalt District, some high grade veins averaged 14 000 to 34 000 or more g Ag/tonne but most deposits were of lower grade. At Great Bear Lake, the mined tonnages and grades to the end of 1978 for the Echo Bay and Terra Mines were, respectively, 393 000 tonnes averaging 2100 g Ag/tonne and 340 000 tonnes averaging 1100 g Ag/tonne.
GEOLOGICAL SETTING	This deposit type can be related in many cases to tensional faults and mafic magmatism. In the Cobalt District, Archean greenstone basement is unconformably overlain by flat-lying Early Aphebian sedimentary rocks and both are intruded by a major Nipissing diabase sheet. In the Great Bear Lake district, Aphebian felsic and intermediate volcaniclastic rocks host the ores. These felsic volcanic rocks mantle a large parental batholith that has invaded its own volcanic cap, and are cut by diabase sills and dykes.
HOST ROCKS OR MINERALIZED ROCKS	Cobalt: Archean mafic volcanic rocks and thin interflow sedimentary units containing sulphidic beds; Aphebian siltstone and greywacke; diabase. Great Bear Lake: waterlain volcaniclastic sedimentary rocks and associated subaerial felsic pyroclastic rocks, in some cases within or near sulphide-rich units in these sequences.
ASSOCIATED ROCKS	Cobalt: Matachewan diabase dykes Great Bear Lake: diabase sill (Echo Bay); alternating red and green tuffaceous rocks; red sandstone, mudstone; syenitic, granodioritic intrusions; magnetite-apatite intrusive bodies.
FORM OF DEPOSIT, DISTRIBUTION OF ORE MINERALS	Steeply dipping veins, roughly perpendicular to associated unconformities and diabase sheets. Cobalt: ore veins occur within about 200 m vertically of the Nipissing diabase sheet and are strongly compositionally zoned; the arsenides grade from Ni through Co to Fe, with Ni arsenides nearest the centre of the diabase sheet; the richest silver ore is associated with the intermediate Ni-Co and Co arsenides. Veins are surrounded successively outward by albite-chlorite, dolomite-calcite-(epidote), and sericite alteration assemblages. Great Bear Lake: at the Terra Mine, steeply dipping veins cut and are roughly perpendicular to a subvertical concordant pyrite-chalcopyrite-rich unit, and extend into the country rocks on either side. Microcline, hematite, carbonates, and possibly chlorite and sericite, are the main alteration minerals. Bleached halos around veins may represent mafic mineral destruction and carbonatization.
MINERALS: Principal ore minerals – ***Associated minerals***	Native silver, acanthite, Ag sulphosalts, pitchblende, Co-Ni arsenides, chalcopyrite. – *Other arsenide minerals, native bismuth, dolomite, calcite, chlorite, quartz, hematite, actinolite*
AGE, HOST ROCKS	Archean, Aphebian.
AGE, ORE	Possibly same as nearby mafic intrusions: 2.1 to possibly 2.3 Ga at Cobalt; 1.39 ± 0.05 Ga at Great Bear Lake.
GENETIC MODEL	Proposed sources for the metals include 1) the older sulphide-bearing units (Cobalt: Archean interflow and basal Aphebian sedimentary rocks; Great Bear Lake: concordant sulphide-rich layers in a felsic pyroclastic sequence), 2) felsic pyroclastic rocks in the case of the Great Bear Lake veins, and 3) the diabase (gabbro) intrusions. The diabase sheets are commonly considered to have given rise to circulating hydrothermal systems and resultant mineralization. While an empirical association with unconformities has been recognized, the genetic significance of this association, if any, has not yet been established.
ORE CONTROLS, GUIDES TO EXPLORATION	1) Unconformities between Archean and Aphebian, or between Aphebian and Helikian sequences. 2) Proximity to diabase sills or sheets. 3) Pre-ore sulphides in nearby rocks (possible source of metals).
AUTHOR	R.I. Thorpe

23. VEIN URANIUM

COMMODITIES	U(Ti, V, Se)
EXAMPLES: Canadian – *Foreign*	Fay-Verna and Gunnar Mines, Beaverlodge area, Sask.; Rayrock Mine, N.W.T.
IMPORTANCE	Canada: significant production in the past (1952-1982) World: constitutes about 6% of current measured plus indicated reserves.
TYPICAL GRADE, TONNAGE	At Beaverlodge up to 13 million tonnes of ore (22 000 tonnes of contained U) in numerous shoots averaging from 0.08 to 0.5% U.
GEOLOGICAL SETTING	In uraniferous metasedimentary and metavolcanic (?) successions, extensively granitized, widely altered (hematite, chlorite, carbonate) and much faulted. At Beaverlodge, close to major faults in rocks that have been mylonitized and closely fractured; at Rayrock, associated with quartz stockwork.
HOST ROCKS OR MINERALIZED ROCKS	At Beaverlodge, mylonitized feldspathic quartzite, brecciated and mylonitized granitic gneiss, altered argillite, and brecciated feldspar-carbonate rocks. At Rayrock, fractured quartz veins in granitized rocks and metasedimentary rocks.
ASSOCIATED ROCKS	At Beaverlodge, basaltic tuff, argillite and granitic gneiss (feldspathic quartzite). At Rayrock, metasedimentary rocks and granite.
FORM OF DEPOSIT, DISTRIBUTION OF ORE MINERALS	At Beaverlodge, veins and tabular bodies concentrated within a specific stratigraphic interval, and elongated parallel to fold plunge. Ore minerals are distributed as fracture-fillings and adjacent disseminations. At Rayrock, as late, uranium-bearing quartz veins.
MINERALS: Principal ore minerals – ***Associated minerals***	Pitchblende; also brannerite at Beaverlodge. – *Hematite, chlorite, carbonate (mainly calcite), quartz*
AGE, HOST ROCKS	Archean (Rayrock); Proterozoic (Beaverlodge).
AGE, ORE	Epigenetic; at Beaverlodge, Hudsonian (1.78 ± 0.02 Ga) with important remobilization in Grenvillian (1.14 ± 0.05 Ga) and at later periods. At Rayrock, remobilization (?) at about 500 Ma, primary age unknown.
GENETIC MODEL	Uranium, extracted possibly from the quartzite and argillite of the metasedimentary-metavolcanic Fay Complex at Beaverlodge and from granitized metasediments at Rayrock, may have been concentrated several times in various environments; was transported hydrothermally; and was deposited in fractures in the mylonitized zones in metasedimentary rocks, granitic gneiss and quartz veins, accompanied by hematite-carbonate alteration. This process is considered to be part of the last stages of metamorphism and to be associated with retrograde effects.
ORE CONTROLS, GUIDES TO EXPLORATION	1. Major faults (represented by wide mylonite zones at Beaverlodge). 2. Late fracturing. 3. Metasedimentary-metavolcanic succession with higher than average uranium content, uraniferous granite and granitized rocks. 4. Areas highly altered to a hematite-chlorite-carbonate assemblage, resulting in pervasively red-coloured rocks. 5. Highly deformed, strongly metamorphosed, granitized terrane.
AUTHORS	L.P. Tremblay, V. Ruzicka

(see Plate 14, page 12)

24. VEIN COPPER

COMMODITIES	Cu (Au, Ag)
EXAMPLES: Canadian – *Foreign*	Opemiska and Chibougamau Districts and Icon area, Que.; Bruce Mines area, Ont.; Churchill area, B.C. – *Magma, Arizona; Morococha, Peru*
IMPORTANCE	Canada: veins and closely associated replacement deposits account for approximately 3% of Canada's copper production and less than 2% of Canada's reserves. World: historically, veins were important in some parts of the world but their present relative importance is greatly diminished.
TYPICAL GRADE, TONNAGE	Highly variable. 1 to 10% Cu, nil to 3 g Au/tonne, nil to 300 g Ag/tonne. 50 000 to 5 million tonnes ore; a few larger deposits are known.
GEOLOGICAL SETTING	Diverse tectonic environments: - Some copper vein deposits are characteristic of hydrothermal districts in continental and island arc settings (in some areas, the copper veins are related to porphyry deposits, e.g., Magma, Arizona). - Deposits in the Opemiska and Chibougamau districts occur in the Archean Abitibi greenstone belt, mainly in differentiated mafic intrusions. The veins are probably genetically related to porphyry deposits and felsic intrusions. - Unusually persistent copper veins and associated diabase bodies occur in tensional settings in Proterozoic sedimentary basins (e.g. Churchill area, B.C. and Bruce Mines area, Ont.); this type of copper vein is widespread in Canada but has been of little economic importance.
HOST ROCKS OR MINERALIZED ROCKS	Diverse; many different types depending on the particular geological settings of the veins (e.g. layered mafic intrusions at Opemiska and Chibougamau, Que.; clastic sedimentary rocks and diabase at Churchill, B.C. and Bruce Mines, Ont.; Precambrian and Paleozoic schist, diabase, quartzite, shale and limestone at Magma, Arizona).
ASSOCIATED ROCKS	Diverse. In some localities in or near mafic, intermediate or felsic intrusive rocks. Hydrothermal alteration (e.g. quartz, feldspar, biotite, sericite, kaolin, chlorite, epidote, calcite, pyrite) of wall rocks is characteristic.
FORM OF DEPOSIT, DISTRIBUTION OF ORE MINERALS	Simple, anastomosing and reticulate veins; vein sets; vein breccia; local stockworks and horsetails. Structurally-controlled ore shoots within veins, especially where the veins change attitude. Patchy, irregular, massive to disseminated sulphides in or near veins. In some areas extensive stringer zones, with or without prominently zoned sulphide and gangue minerals, are common.
MINERALS: Principal ore minerals – *Associated minerals*	Chalcopyrite, bornite, chalcocite, tetrahedrite, native gold. *– Pyrite, pyrrhotite, other sulphides and sulphosalts, hematite, magnetite, quartz, calcite, dolomite, ankerite, siderite, chlorite, sericite*
AGE, HOST ROCKS	Archean to Tertiary.
AGE, ORE	Archean to Tertiary. Ores are epigenetic and may be much younger than the host rocks.
GENETIC MODEL	Hydrothermal ore deposition along fissures and faults. In some localities probably related to emplacement of mafic, intermediate or felsic intrusions.
ORE CONTROLS, GUIDES TO EXPLORATION	1. Rifted Proterozoic sedimentary successions may contain sizeable copper veins. 2. Sulphides may be localized within the parts of veins that cross-cut carbonates or other favourable wall rocks. 3. Change in attitude of vein may localize high grade shoots.
AUTHOR	R.V. Kirkham

25. FELSIC INTRUSION—ASSOCIATED SILVER—LEAD—ZINC VEINS

COMMODITIES	Ag, Pb, Zn (Cd, Au, Cu)
EXAMPLES: Canadian – ***Foreign***	Canada: of the numerous Ag-Pb-Zn vein deposits known in Canada, the most important occur in the Slocan-New Denver-Ainsworth District, southeastern B.C. (e.g. Highland Bell, Lucky Jim) and the Keno Hill-Galena Hill District, Yukon (e.g. Hector-Calumet, Elsa).
IMPORTANCE	Canada: historically a major source of Ag, Pb, and to a lesser extent, Zn; presently of minor and diminishing importance.
TYPICAL GRADE, TONNAGE	Canadian deposits consist of individual veins or vein-systems that range from 10 tonnes to two million tonnes with a modal size of about 100 000 tonnes. Metal grades show a wide range depending to some degree on the extent of selective mining and/or sorting in the earlier-mined deposits. Excluding the more obvious hand-sorted deposits, grades in Canadian deposits of this type range as follows: 5 to 1500 g Ag/tonne, 0.5 to 20% Pb, 0.5 to 8% Zn. Typical grades are: 300 to 900 g Ag/tonne, 4 to 8% Pb, 2 to 4% Zn.
GEOLOGICAL SETTING	Largely in thick sequences of clastic sedimentary rocks (shale, siltstone, quartzite); nearby igneous intrusions, usually felsic in composition.
HOST ROCKS OR MINERALIZED ROCKS	Quartzite, siltstone, shale, carbonates, igneous intrusions.
ASSOCIATED ROCKS	Felsic stocks or plutons; less commonly gabbro or diabase.
FORM OF DEPOSIT, DISTRIBUTION OF ORE MINERALS	Most commonly discordant, narrow, elongate, steeply-dipping zones. Some offset of wall rocks usually apparent so zones are best described as vein-faults. Clusters of parallel veins/vein-faults referred to as vein-sets. In any one district, one particular structural trend usually carries the mineralization while other trends in the same district do not. Ore minerals largely sharply confined within vein or vein-fault; wall rock replacement and disseminations also occur but are not common. Vein minerals display cockade texture, cement angular fragments of wall-rock, or occur as coarse grained masses.
MINERALS: Principal ore minerals – ***Associated minerals***	Galena, sphalerite; silver is contained in galena as distinct minerals such as native silver, argentite, freibergite, and various sulphosalt minerals, most commonly tetrahedrite. – *Pyrite, chalcopyrite, arsenopyrite; quartz, carbonates (siderite, ankerite, calcite), barite, fluorite*
AGE, HOST ROCKS	Helikian to Cretaceous (?) in Canada. Rocks of Mesozoic age are probably the most common hosts in Canada.
AGE, ORE	Unknown; obviously younger than host rocks and, in most instances, can be shown to be younger than nearby felsic intrusions. Age of mineralization is difficult to determine because deposits typically do not contain minerals amenable to radiometric dating.
GENETIC MODEL	Hydrothermal open-space filling, with minor replacement, along fractures or dilatant zones. These fracture zones may be genetically related to nearby intrusions. However, this type of deposit has received relatively little research, and no generally accepted genetic model has emerged.
ORE CONTROLS, GUIDES TO EXPLORATION	Fractures and faults in thick, clastic sedimentary sequences invaded by felsic intrusions.
AUTHOR	D.F. Sangster

26. ULTRAMAFIC—HOSTED ASBESTOS

COMMODITIES	Chrysotile
EXAMPLES: Canadian – ***Foreign***	Jeffrey Mine, Asbestos, Que.; Bell-King-Beaver Mines, Thetford Mines, Que.; British Canadian Mine, Black Lake, Que.; Asbestos Hill, Ungava, Que.; Advocate Mine, Baie Verte, Nfld.; Cassiar Mine, B.C.; Munro Mine, Matheson, Ont.; Clinton Creek, Yukon – *Bazhenov, U.S.S.R.; Msauli Mine, South Africa; Havelock Mine, Swaziland; King Mine, Mashaba, Zimbabwe*
IMPORTANCE	Chrysotile asbestos accounts for nearly 5% of the total value of Canadian mineral production (excluding petroleum and natural gas). Canada accounts for about 30% of world asbestos production, ranking second to the U.S.S.R.
TYPICAL GRADE, TONNAGE	Of the order of 10 to 100 million tonnes containing 2 to 10% recoverable fibre. Examples (past production plus reserves): Jeffrey, Bell-King-Beaver, and British Canadian Mines, respectively, about 800 million, 250 million and 150 million tonnes averaging about 6% fibre; Advocate Mine, about 60 million tonnes averaging about 3% recoverable fibre; and Cassiar, about 23 million tonnes with fibre recoveries averaging about 7 to 10%.
GEOLOGICAL SETTING	(1) The most important deposits occur in allochthonous bodies of serpentinized ophiolitic or alpine ultramafic rocks in Phanerozoic orogenic belts (e.g., the deposits in the Appalachians, Cordillera, and Urals). (2) Deposits of lesser importance occur in the ultramafic zones of synvolcanic intrusions in Precambrian greenstone belts (e.g. the deposits of northeastern Ontario, Ungava, Quebec, and southern Africa).
HOST ROCKS OR MINERALIZED ROCKS	Serpentinized, olivine-rich ultramafic rocks, most commonly harzburgite, but also dunite, wehrlite, and orthopyroxenite.
ASSOCIATED ROCKS	Serpentinized peridotite, pyroxenite, steatite (talc-magnesite rock), gabbro, basalt, rodingite. Granitic and other felsic bodies occur in association with deposits in southern Quebec.
FORM OF DEPOSIT, DISTRIBUTION OF ORE MINERALS	Orebodies are equidimensional to somewhat oblate zones, of the order of 100 to 1000 m on a side, within masses of serpentinized ultramafic rock. The ore typically comprises a stockwork of chrysotile veins. Individual veins are for the most part less than about 1 cm thick but may be up to about 10 cm thick and several metres in length. Cross fibre veins where the chrysotile fibres are at a high angle to the vein walls are more abundant than slip fibre veins in which the fibres are oriented along the length of the vein.
MINERALS: Principal ore minerals – ***Associated minerals***	Chrysotile. – *Magnetite, brucite, antigorite, lizardite, talc, carbonate, chromite; nephrite jade is a minor byproduct of asbestos production at Cassiar*
AGE, HOST ROCKS	Archean (Abitibi area, Ontario and Quebec; southern Africa), Aphebian (Ungava area, Quebec), lower Paleozoic (southern Quebec; Newfoundland), upper Paleozoic (Cassiar, British Columbia).
AGE, ORE	Corresponds to age of deformation and metamorphism of host rocks.
GENETIC MODEL	Chrysotile is deposited in fractures during deformation of ultramafic rock under relatively low grade metamorphic conditions (greenschist or subgreenschist). The predominance of cross fibre veins indicates fracturing under tensional stress or dilation of pre-existing fractures. The vein-filling material (mainly chrysotile, brucite, and magnetite) forms from constituents derived locally, possibly even by lateral secretion.
ORE CONTROLS, GUIDES TO EXPLORATION	1. Presence of serpentinized ultramafic rock, preferably harzburgite or dunite. 2. Evidence of brittle as opposed to ductile deformation: abundant faults, fractures and tension gashes.
AUTHOR	J.M. Duke (see Plate 15, page 12)

SELECTED BIBLIOGRAPHY

1 EVAPORITES AND BRINES

Davis, J.R. and Vine, J.D.
1979: Stratigraphic and tectonic setting of the lithium brine field, Clayton Valley, Nevada; Rocky Mountain Association of Geologists – Utah Geological Association 1979 Basin and Range Symposium, ed., G.W. Newman and H.D. Good, p. 421-430.

Gwynn, J.W., editor
1980: Great Salt Lake – a scientific, historical and economic overview; Utah Department of Natural Resources, Bulletin 116, 400 p.

Johnson, K.S. and Gonzales, S.
1978: Salt deposits in the United States and regional geologic characteristics important for storage of radioactive waste; prepared for The Office of Waste Isolation, Union Carbide Corporation, Nuclear Division, United States Department of Energy; published by Earth Resource Associates, Inc., Athens, Georgia, 188 p.

Kendall, A.C.
1978: Facies Models-12. Subaqueous evaporites; Geoscience Canada, v. 5, no. 3, p. 124-129.

Lefond, S.J., editor
1975: Industrial minerals and rocks (fourth edition); American Institute of Mining, Metallurgical and Petroleum Engineers, Inc., New York, 1360 p.

Mann, A.W. and Deutscher, R.L.
1978: Genesis principles for the precipitation of carnotite in calcrete drainages in Western Australia; Economic Geology, v. 73, p. 1724-1737.

Nissenbaum, A., editor
1980: Hypersaline brines and evaporitic environments; Developments in Sedimentology: 28; Elsevier Scientific Publishing Company, Amsterdam-Oxford-New York, 270 p.

Skinner, B.J., editor
1979: An issue devoted to phosphate, potash and sulfur; Economic Geology, v. 74, 304 p.

Smith, G.I., Jones, C.L., Culbertson, W.C., Ericksen, G.E. and Dyni, J.R.
1973: Evaporites and brines; in United States Mineral Resources, ed., D.A. Brobst and W.P. Pratt; United States Geological Survey, Professional Paper 820, p. 197-216.

2 IRON-RICH SEDIMENTARY STRATA

2.1 Ironstone

Gross, G.A.
1970: Nature and occurrence of iron ore deposits; in Survey of World Iron Ore Resources; United Nations Publication, p. 13-31.

1970: Iron ore deposits of Canada and the West Indies; in Survey of World Iron Ore Resources; United Nations Publication, p. 237-269.

Hayes, A.O.
1915: Wabana Iron Ore of Newfoundland; Geological Survey of Canada, Memoir 78, 163 p.

Mellon, G.B.
1962: Petrology of Upper Cretaceous oolitic iron-rich rocks from northern Alberta; Economic Geology, v. 57, p. 921-940.

2.2 Iron-Formation (Lake Superior Type)

Gross, G.A.
1965: Geology of Iron Deposits in Canada. Volume I. General Geology and Evaluation of Iron Deposits; Geological Survey of Canada, Economic Geology Report 22, 181 p.

1968: Geology of Iron Deposits in Canada. Volume III. Iron Ranges of the Labrador Geosyncline; Geological Survey of Canada, Economic Geology Report 22, 179 p.

1970: Nature and occurrence of iron ore deposits: Iron ore deposits of Canada and the West Indies; in Survey of World Iron Ore Resources; United Nations Publication, p. 13-31, 237-269.

James, H.L. and Sims, P.K., co-editors
1973: Precambrian Iron-Formations of the World; Economic Geology, v. 68, no. 7, p. 913-1179.

Klemic, H.
1970: Iron ore deposits of the United States of America, Puerto Rico, Mexico and Central America; in Survey of World Iron Ore Resources; United Nations Publication, p. 411-477.

Zajac, I.S.
1974: The Stratigraphy and Mineralogy of the Sokoman Formation in the Knob Lake Area, Quebec and Newfoundland; Geological Survey of Canada, Bulletin 220, 159 p.

2.3 Iron-Formation (Algoma Type)

Gross, G.A.
1965: Geology of iron deposits in Canada. Volume I. General geology and evaluation of iron deposits; Geological Survey of Canada, Economic Geology Report 22, 181 p.

1973: The depositional environment of principal types of Precambrian iron-formation; in Genesis of Precambrian Iron and Manganese Deposits; Proceedings Kiev Symposium, UNESCO Earth Sciences 9, p. 15-21.

1980: A classification of iron-formations based on depositional environments; Canadian Mineralogist, v. 18, p. 215-222.

3 ENRICHED IRON-FORMATION

Ayers, D.E.
1971: The hematite ores of Mount Tom Price and Mount Whaleback Hamersley iron province; Proceedings Australasian Institute of Mining and Metallurgy, no. 238, p. 47-58.

Belevtsev, Y.N.
1973: Genesis of high-grade iron ores of Krivoyrog Type; in Genesis of Precambrian Iron and Manganese Deposits; Proceedings of the Kiev Symposium, UNESCO Earth Sciences 9, p. 167-180.

Gross, G.A.
1965: Geology of Iron Deposits in Canada. Volume I. General geology and evaluation of iron deposits; Geological Survey of Canada, Economic Geology Report 22, 181 p.

1968: Geology of Iron Deposits in Canada. Volume III. Iron ranges of the Labrador Geosyncline; Geological Survey of Canada, Economic Geology Report 22, 179 p.

Joliffe, A.W.
1966: Stratigraphy of the Steeprock Group, Steep Rock Lake, Ontario; in Precambrian Symposium; Geological Association of Canada, Special Paper 3, p. 75-98.

Stubbins, J.B., Blais, R.A. and Zajac, S.I.
1961: Origin of the soft iron ores of the Knob Lake Range; Transactions Canadian Institute of Mining and Metallurgy, v. 64, p. 37-52.

United Nations, Department of Economic and Social Affairs
1970: Survey of World Iron Ore Resources, United Nations Publication, 479 p.

4 STRATIFORM PHOSPHATE (PHOSPHORITE)

Bentor, Y.K.
1980: Phosphorites – the unsolved problems; in Marine Phosphorites – Geochemistry, Occurrence, Genesis, ed. Y.K. Bentor; Society of Economic Paleontologists and Mineralogists, Special Publication, no. 29, p. 3-18.

British Sulphur Corporation
1980: World Survey of Phosphate Deposits, 4th edition; British Sulphur Corporation Limited, London, 238 p.

Cathcart, J.B.
1980: World phosphate reserves and resources; in The Role of Phosphorus in Agriculture, ed. F.E. Khasawnch et al.; American Society of Agronomy, Inc., Crop Science Society of America, Inc., Soil Science Society of America, Inc., p. 1-18.

Christie, R.L.
1978: Sedimentary phosphate deposits – an interim review; Geological Survey of Canada, Paper 78-20, 9 p.

Cook, P.J.
1976: Sedimentary phosphate deposits; in Handbook of Strata-bound and Stratiform Ore Deposits, Volume 7, ed. K.H. Wolf, Elsevier, Amsterdam, p. 505-535.

Notholt, A.J.G.
1980: Economic phosphatic sediments: mode of occurrence and stratigraphical distribution; Journal of the Geological Society of London, v. 137, no. 6, p. 793-805.

Notholt, A.J.G., Highley, D.E., and Slansky, M.
1979: Dossier on phosphate; Dossier IV of Raw Materials Research and Development; Commission of the European Communities, DG XIII – Research, Science, Education.

Sheldon, R.P.
1981: Ancient marine phosphorites; Annual Review of Earth and Planetary Sciences, v. 9, p. 251-284.

Slansky, M.
1980: Géologie des phosphates sédimentaires; Bureau de Recherches Géologiques et Minières, France, Mémoire no. 114, 92 p.

5 PLACER URANIUM, GOLD

5.1 Pyritic Paleoplacer Uranium, Gold

Armstrong, F.C., ed.
1981: Genesis of uranium- and gold-bearing Precambrian quartz-pebble conglomerates; United States Geological Survey, Professional Paper 1161A-BB.

Pretorius, D.A.
1981: Gold and uranium in quartz-pebble conglomerates; University of the Witwatersrand, Economic Geology Research Unit, Information Circular No. 151.

Roscoe, S.M.
1969: Huronian rocks and uraniferous conglomerates in the Canadian Shield; Geological Survey of Canada, Paper 68-40, 205 p.

5.2 Placer Gold

Boyle, R.W.
1979: The geochemistry of gold and its deposits; Geological Survey of Canada, Bulletin 280, 584 p.

Lay, D.
1941: Fraser River Tertiary drainage-history in relation to placer-gold deposits (Part II); British Columbia Department of Mines, Bulletin 11, 75 p.

McConnell, R.E.
1905: Report on the Klondike gold fields; Geological Survey of Canada Annual Report for 1901 (New Series), Volume XIV, p. 5B-71B. Reprinted in Geological Survey of Canada Memoir 284, p. 64-113.

6 STRATABOUND SEDIMENT-HOSTED LEAD, ZINC, COPPER, URANIUM

6.1 Mississippi Valley Lead-Zinc

Beales, F.W. and Jackson, S.A.
1966: Precipitation of lead-zinc ores in carbonate reservoirs as illustrated by Pine Point ore field, Canada; Transactions, Institute of Mining and Metallurgy (London), Section B, v. 75, p. B278-285.

Economic Geology
1971: A paleoacquifer and its relation to economic mineral deposits: The Lower Ordovician Kingsport Formation and Mascot Dolomite: A symposium; Economic Geology, v. 66, no. 5, p. 695-810.

Hagni, R.D.
1976: Tri-State ore deposits: The character of their host rocks and their genesis; in Handbook of Strata-bound and Stratiform Ore Deposits Volume 6, ed. K. Wolf; Elsevier Scientific Publishing Company, Amsterdam, p. 457-494.

Heyl, A.V.
1968: The Upper Mississippi Valley base-metal district; in Ore Deposits of the United States, Volume 1, ed. J.R. Ridge; American Institute of Mining, Metallurgical, and Petroleum Engineers, Inc., New York, p. 431-459.

Hoagland, A.D.
1976: Appalachian zinc-lead deposits; in Handbook of Strata-bound and Stratiform Ore Deposits Volume 6, ed. K. Wolf; Elsevier Scientific Publishing Co., Amsterdam, p. 495-534.

Kyle, J.R.
1977: Development of sulfide-hosting structures and mineralization, Pine Point, Northwest Territories; unpublished Ph.D. thesis, University of Western Ontario, London, 226 p.

Ohle, E.L.
1980: Some considerations in determining the origin of ore deposits of the Mississippi Valley type-Part II; Economic Geology, v. 75, p. 161-172.

Skall, H.
1975: The paleoenvironment of the Pine Point lead-zinc district; Economic Geology, v. 70, p. 22-45.

Vineyard, J.D., ed.
1977: An issue devoted to the Viburnum Trend, southeast Missouri; Economic Geology, v. 72, no. 3, p. 337-486.

6.2 Sandstone Lead

Bjørlykke, A. and Sangster, D.F.
1981: An overview of sandstone-lead deposits and their relation to red-bed copper and carbonate-hosted lead-zinc deposits; in Economic Geology 75th Anniversary Volume, 1905-1980, ed. B.J. Skinner; Economic Geology Publishing Co., p. 179-213.

Fogliérini, F., Samama, J.C., and Rey, M.
1980: Le gisement stratiforme de Largentière (ardèche); Bureau de Recherches Géologiques et Minières, Memoire 112-E4, 54 p.

Rickard, D.T., Willdén, M.Y., Marinder, N.E., and Donnelly, T.H.
1979: Studies on genesis of the Laisvall sandstone lead-zinc deposit, Sweden; Economic Geology, v. 74, p. 1255-1285.

Samama, J.C.
1976: Comparative review of the genesis of the copper-lead sandstone-type deposits; in Handbook of Strata-bound and Stratiform Ore Deposits, Volume 6, ed. K. Wolf; Elsevier Scientific Publishing Company, Amsterdam, p. 1-20.

6.3 Sedimentary Copper

Annels, A.E.
1979: The genetic relevance of recent studies at Mufulira Mine, Zambia; Annales de la Société Géologique de Belgique, v. 102, p. 431-449.

Renfro, A.R.
1974: Genesis of evaporite-associated stratiform metalliferous deposits – a sabkha process; Economic Geology, v. 69, p. 33-45.

Rentzsch, J.
1974: The Kupferschiefer in comparison with the deposits of the Zambian Copperbelt; in Gisements Stratiformes et Provinces Cuprifères, ed. P. Bartholomé; Centenaire de la Société Géologique de Belgique, Liege, p. 395-418.

Rose, A.W.
1976: The effect of cuprous chloride complexes in the origin of red bed copper and related deposits; Economic Geology, v. 71, p. 1036-1048.

Smith, G.E.
1976: Sabkha and tidal-flat facies control of stratiform copper deposits in north Texas; in Stratiform Copper Deposits of the Midcontinent Region, a Symposium, ed. K.S. Johnson and R.L. Croy; Oklahoma Geological Survey, Circular 77, p. 25-39.

Strakhov, N.M.
1962: Principles of lithogenesis; translated by J.P. Fitzsimmons, 1970; ed. S.I. Tomkeieff and J.E. Hemingway, Plenum Publishing Corporation, New York and Oliver & Boyd, Edinburgh, v. 3, 577 p.

6.4 Sandstone Uranium

Danchev, V.I. and Lapinskaya, T.A.
1965: Deposits of Radioactive Raw Material; Nedra Press, Moscow (in Russian).

Finch, W.I.
1967: Geology of Epigenetic Uranium Deposits in sandstones in the United States; United States Geological Survey Professional Paper 538, 121 p.

Gabelman, J.W.
1971: Migration of uranium and thorium – exploration significance; American Association of Petroleum Geologists, Studies in Geology, no. 3, 168 p.

International Atomic Energy Agency, Vienna
1974: Formation of Uranium Ore Deposits; Proceedings of a Symposium on the formation of Uranium ore deposits, Athens, Greece, 1974, 748 p.

Langford, F.F.
1977: Surficial origin of North American pitchblende and related uranium deposits; American Association of Petroleum Geologists, Bulletin, v. 61, p. 28-42.

7 CHEMICAL-SEDIMENT-HOSTED GOLD

Barnett, E.S., Hutchinson, R.W., Adamcik, A., and Barnett, R.
1982: Geology of the Agnico-Eagle gold deposit, Quebec; in Precambrian Sulphide Deposits, H.S. Robinson Memorial Volume, ed. R.W. Hutchinson, C.D. Spence and J.M. Franklin; Geological Association of Canada, Special Paper 25, p. 403-426.

Fleischer, R. and Routhier, P.
1973: The "consanguineous" origin of a tourmaline-bearing gold deposit: Passagem de Mariana (Brazil); Economic Geology, v. 68, p. 11-22.

Gibbins, W.
Gold and Precambrian iron formation in the Northwest Territories; Department of Indian Affairs and Northern Development, Yellowknife, N.W.T. (in press)

McConnell, G.W.
1964: Notes on similarities between some Canadian gold deposits and the Homestake deposits of South Dakota; Economic Geology, v. 59, p. 719-720.

Moreschi, J.B.
1977: A mina du ouro do Faria, Minas Gerais – um deposito estratiforme associado a um complexo vulcano-sedimentar; Universidade de Sao Paulo, Instituto de Geosciências, Boletim, v. 8, p. 119-138.

Sawkins, F.J. and Rye, D.M.
1974: Relationship of Homestake type gold deposits to iron-rich Precambrian sedimentary rocks; Transactions Institute of Mining and Metallurgy, Section B (Applied Earth Science), v. 83, no. 2, p. 56-59.

8 CLASTIC-SEDIMENT-HOSTED GOLD

8.1 Carbonaceous Shale/Carbonate-Hosted Gold (Carlin Type)

Dickson, F.W., Rye, R.O. and Radtke, A.S.
1979: The Carlin gold deposit as a product of rock-water interactions; in Papers on Mineral Deposits of Western North America, ed. J.D. Ridge; Nevada Bureau of Mines and Geology, Report 33, p. 101-108.

Joralemon, P.
1978: A major gold belt takes shape in Nevada; Mining Engineering, July 1978, p. 750-762.

Radtke, A.S., Rye, R.O. and Dickson, F.W.
1980: Geology and stable isotope studies of the Carlin gold deposit, Nevada; Economic Geology, v. 75, p. 641-672.

Silberman, M.L., Berger, R.B. and Koski, R.A.
1974: K-Ar relations of granodiorite emplacement and tungsten and gold mineralization near the Getchell Mine, Humboldt County, Nevada; Economic Geology, v. 69, p. 646-656.

Wells, J.D., Stoiser, L.R. and Elliott, J.E.
1969: Geology and geochemistry of the Cortez gold deposit, Nevada; Economic Geology, v. 64, p. 526-537.

8.2 Turbidite-Hosted Vein and Shear Zone Gold

Chace, F.M.
1949: Origin of the Bendigo saddle reefs and formation of ribbon quartz; Economic Geology, v. 44, p. 561-597.

Graves, M.C.
1976: The formation of gold-bearing quartz veins in Nova Scotia: Hydraulic fracturing under conditions of greenschist regional metamorphism during early stages of deformation; M.Sc. thesis, Dalhousie University, Halifax, 158 p.

Malcolm, W.
1976: Gold Fields of Nova Scotia; Geological Survey of Canada, Memoir 385 (Reprinting of Memoir 156 published in 1929), 253 p.

9 STRATIFORM SULPHIDE, BARITE

9.1 Volcanic-Associated Massive Sulphide

Constantinou, G.
1980: Metallogenesis associated with the Troodos Ophiolite; in Ophiolites: International Ophiolite Symposium, Cyprus, 1979, Proceedings; Cyprus Ministry of Agriculture and Natural Resources, Geological Survey Department, p. 663-674.

Franklin, J.M., Lydon, J.W. and Sangster, D.F.
1981: Volcanic-associated massive sulfide deposits; in Economic Geology 75th Anniversary Volume, 1905-1980, ed. B.J. Skinner; Economic Geology Publishing Co., p. 485-627.

Klau, W. and Large, D.E.
1980: Submarine exhalative Cu-Pb-Zn deposits – A discussion of their classification and metallogeneis; Geologisches Jahrbuch, series D, no. 40, p. 13-58.

Lambert, I.B. and Sato, T.
1974: The Kuroko and associated ore deposits of Japan: A review of their features and metallogenesis; Economic Geology, v. 69, p. 1215-1236.

Sangster, D.F.
1972: Precambrian volcanogenic massive sulphide deposits in Canada: A review; Geological Survey of Canada, Paper 72-22, 44 p.

Sangster, D.F. and Scott, S.D.
1976: Precambrian, strata-bound, massive Cu-Zn-Pb sulfide ores of North America; in Handbook of Strata-bound and Stratiform Ore Deposits, Volume 6, ed. K.H. Wolf; Elsevier Scientific Publishing Company, Amsterdam, p. 129-222.

Solomon, M.
1976: "Volcanic" massive sulphide deposits and their host rocks – a review and an explanation; in Handbook of Strata-bound and Stratiform Ore Deposits, Volume 6, ed. K.A. Wolf; Elsevier Scientific Publishing Company, Amsterdam, p. 21-50.

Swanson, E.A., Strong, D.F. and Thurlow, J.G., ed.
1981: The Buchans Ore Bodies: Fifty Years of Geology and Mining; Geological Association of Canada, Special Paper 22, 350 p.

9.2 Sediment-Hosted Sulphide

Gustafson, L.B., and Williams, N.
1981: Sediment-hosted stratiform deposits of copper, lead, and zinc; in Economic Geology Seventy-fifth Anniversary Volume, 1905-1980, ed. B.J. Skinner; Economic Geology Publishing Co., p. 139-178.

Hamilton, J.M., Bishop, D.T., Morris, H.C., and Owens, O.E.
1982: Geology of the Sullivan orebody, Kimberly, B.C., Canada; in Precambrian Sulphide Deposits, H.S. Robinson Memorial Volume, ed. R.W. Hutchinson, C.D. Spence, and J.M. Franklin; Geological Association of Canada, Special Paper 25, p. 597-665.

Hannak, W.W.
1981: Geology of the Rammelsberg ore deposit near Goslar/Upper Harz, Federal Republic of Germany; in Handbook of Strata-bound and Stratiform Ore Deposits, Volume 9, ed. K.H. Wolf; Elsevier Scientific Publishing Company, Amsterdam, p. 551-642.

Krebs, W.
1981: The geology of the Meggen ore deposit; in Handbook of Strata-bound and Stratiform Ore Deposits, Volume 9, ed. K.H. Wolf; Elsevier Scientific Publishing Company, Amsterdam, p. 509-549.

Large, D.E.
1980: Geological parameters associated with sediment-hosted, submarine exhalative Pb-Zn deposits: an empirical model for mineral exploration; Geologisches Jahrbuch, Series D, no. 40, p. 59-129.

9.3 Sediment-Hosted Barite

Dawson, K.R.
Barium, Strontium and Fluorine Deposits in Canada; Geological Survey of Canada, Economic Geology Report 34 (in preparation)

Rye, R.D., Shawe, D.R. and Poole, F.G.
1978: Stable isotope studies of bedded barite at East Northumberland Canyon in Toquima Range, Central Nevada; Journal of Research, U.S. Geological Survey, v. 6, p. 221-229.

Zimmerman, R.A.
1970: Sedimentary features in the Meggen barite-pyrite-sphalerite deposit and a comparison with the Arkansas barite deposits; Neues Jahrbuch fur Mineralogie, Abhandlungen, v. 113, p. 179-214.

10 VOLCANIC REDBED COPPER

Harper, G.
1977: Geology of the Sustut copper deposit in B.C.: The Canadian Mining and Metallurgy Bulletin, v. 70, no. 777, p. 97-104.

Jolly, W.T.
1974: Behavior of Cu, Zn, and Ni during prehnite-pumpellyite rank metamorphism of the Keweenawan basalts, northern Michigan; Economic Geology, v. 69, p. 1118-1125.

Kirkham, R.V.
1982: Volcanic red bed copper deposits – environments of formation and distribution in accreted terranes of western North America (abstract); in Rocks and Ores of the Middle Ages, Programme and Abstracts; Cordilleran Section, The Geological Association of Canada, p. 14-16.

Lincoln, T.N.
1981: The redistribution of copper during low-grade metamorphism of the Karmutsen volcanics, Vancouver Island, British Columbia; Economic Geology, v. 76, no. 8, p. 2147-2161.

Ruiz, C., Aguilar, A., Egert, E., Espinosa, W., Peebles, F., Quezada, R. and Serrano, M.
1971: Strata-bound copper sulphide deposits of Chile; The Society of Mining Geologists of Japan, Special Issue 3, p. 252-260.

White, W.S.
1968: The native-copper deposits of northern Michigan; in Ore Deposits of the United States, 1933-1967 (Graton-Sales Volume), ed. J.D. Ridge; American Institute of Mining, Metallurgical and Petroleum Engineers, Inc., New York, p. 303-325.

11 VOLCANIC-ASSOCIATED VEIN AND SHEAR ZONE GOLD

Boyle, R.W.
1961: The geology, geochemistry, and origin of the gold deposits of the Yellowknife district; Geological Survey of Canada, Memoir 310, 193 p.

Boyle, R.W.
1979: The geochemistry of gold and its deposits; Geological Survey of Canada, Bulletin 280, 584 p.

Franklin, J.M. and Thorpe, R.I.
1982: Comparative metallogeny of the Superior, Slave and Churchill provinces; in Precambrian Sulphide Deposits, H.S. Robinson Memorial Volume, ed. R.W. Hutchinson, C.D. Spence and J.M. Franklin; Geological Association of Canada, Special Paper 25, p. 3-90.

Karvinen, W.O.
1981: Geology and evolution of gold deposits, Timmins area, Ontario; in Genesis of Archean, Volcanic-Hosted Gold Deposits, Symposium held at the University of Waterloo, March 7, 1980, ed. E.G. Pye and R.G. Roberts; Ontario Geological Survey Miscellaneous Paper 97, p. 29-46.

Kerrich, R., and Fryer, B.J.
1979: Archean precious-metal hydrothermal systems, Dome mine, Abitibi greenstone belt. II. REE and isotopic relations; Canadian Journal of Earth Sciences, v. 16, p. 440-458.

12 MAGMATIC NICKEL, COPPER, PLATINUM GROUP ELEMENTS

12.1 Ultramafic-Associated Nickel-Copper

Groves, D.I., Hudson, D.R., Marston, R.J., and Ross, J.R. (editors)
1981: A special issue on nickel deposits and their host rocks in Western Australia; Economic Geology, v. 76, p. 1289-1783.

Naldrett, A.J., ed.
1979: Nickel-sulfide and platinum group element deposits: Proceedings of an international symposium, October 21-22, 1978; Canadian Mineralogist, v. 17, pt. 2, p. 141-514.

Naldrett, A.J.
1981: Nickel sulfide deposits: classification, composition and genesis; in Economic Geology Seventy-fifth Anniversary Volume, 1905-1980, ed. B.J. Skinner; Economic Geology Publishing Co., p. 628-685.

12.2 Gabbroid-Associated Nickel, Copper, Platinum Group Elements

Glaskovsky, A.A., Gorbunov, G.I., and Sysoev, F.A.
1977: Deposits of nickel; in Ore Deposits of the U.S.S.R., Volume II, ed. V.I. Smirnov (translated into English by D.A. Brown); Pitman Publishing, London, p. 3-79.

Groves, D.I., Hudson, D.R., Marston, R.J., and Ross, J.R. (editors)
1976: An issue devoted to platinum-group elements; Economic Geology, v. 71, no. 7, p. 1129-1480.

Naldrett, A.J. (editor)
1979: Nickel-sulfide and platinum-group-element deposits: Proceedings of an international symposium, October 21-22, 1978; Canadian Mineralogist, v. 17, pt. 2, p. 141-514.

Naldrett, A.J.
1981: Nickel sulphide deposits: classification, composition and genesis; in Economic Geology Seventy-fifth Anniversary Volume, 1905-1980, ed. B.J. Skinner; Economic Geology Publishing Co., p. 628-685.

Pattison, E.F.
1979: The Sudbury Sublayer; Canadian Mineralogist, v. 17, p. 257-274.

Pinsent, R.H.
1980: Nickel-copper mineralization in the Lynn Lake gabbro; Manitoba Mineral Resources Division, Economic Geology Report ER79-3, 138 p.

13 MAFIC/ULTRAMAFIC-HOSTED CHROMITE

13.1 Stratiform

Cameron, E.N. and Desborough, G.A.
1969: Occurrence and characteristics of chromite deposits – Eastern Bushveld Complex; in Magmatic Ore Deposits, ed. H.D.B. Wilson; Economic Geology Monograph 4, p. 23-40.

Jackson, E.D.
1961: Primary textures and mineral associations in the Ultramafic Zone of the Stillwater Complex, Montana; United States Geological Survey Professional Paper 358, 106 p.

Irvine, T.N.
1975: Crystallization sequence in the Muskox intrusion and other layered intrusions – II. Origin of chromitite layers and similar deposits of other magmatic ores; Geochimica et Cosmochimica Acta, v. 39, p. 991-1020.

Irvine, T.N.
1977: Origin of chromitite layer in the Muskox intrusion and other stratiform intrusions: A new interpretation; Geology, v. 5, p. 273-277.

13.2 Podiform

Dickey, J.S., Jr.
1975: A hypothesis of origin for podiform chromite deposits; Geochimica et Cosmochimica Acta, v. 39, p. 1061-74.

Greenbaum, D.
1977: The chromitiferous rocks of the Troodos ophiolite complex, Cyprus; Economic Geology, v. 72, p. 1175-94.

Thayer, T.P.
1964: Geologic features of podiform chromite deposits; in Methods of Prospection for Chromite, ed. R. Woodtli; Organization for Economic Co-Operation and Development, Paris, p. 135-146.

1964: Gravity differentiation and magmatic re-emplacement of podiform chromite deposits; in Magmatic Ore Deposits, ed. H.D.B. Wilson; Economic Geology Monograph 4, p. 132-146.

14 MAFIC INTRUSION-HOSTED TITANIUM-IRON

Allard, G.O.
1976: Doré Lake Complex; Québec Ministère des Richesses Naturelles, DP-368, 446 p.

Gross, G.A.
1965: Geology of Iron Deposits in Canada: Volume I. General Geology and Evaluation of Iron Deposits; Geological Survey of Canada, Economic Geology Report 22, 181 p.

1967: Geology of Iron Deposits in Canada: Volume II. Iron Deposits in the Appalachian and Grenville Regions of Canada; Geological Survey of Canada, Economic Geology Report 22, 111 p.

Hammond, P.
1952: Allard Lake ilmenite deposits; Economic Geology, v. 47, p. 634-649.

Hargraves, R.B.
1962: Petrology of the Allard Lake anorthosite suite, Quebec; in Petrologic Studies: A Volume in Honor of A.F. Buddington, ed. A.E.J. Engel, H.L. James and B.F. Leonard; Geological Society of America, p. 163-190.

Lister, G.F.
1966: The composition and origin of selected iron-titanium deposits; Economic Geology, v. 61, p. 275-310.

Rose, E.R.
1969: Geology of titanium and titaniferous deposits of Canada; Geological Survey of Canada, Economic Geology Report 25, 177 p.

15 INTRUSION-ASSOCIATED GOLD

Bedard, P. and Imbeau, G.
1980: Compagnie Minière Lamaque 1964 Limitée; in The Canadian Institute of Mining and Metallurgy, Geology Division CIM Gold Symposium and Field Excursion, Val d'Or-Kirland Lake-Timmins, ed. W. Petruk; The Canadian Institute of Mining and Metallurgy, p. 52-60.

Franklin, J.M. and Thorpe, R.I.
1982: Comparative metallogeny of the Superior, Slave and Churchill Provinces; in Precambrian Sulphide Deposits, H.S. Robinson Memorial Volume, ed. R.W. Hutchinson, C.D. Spence and J.M. Franklin; Geological Association of Canada, Special Paper 25, p. 3-90.

Fyfe, W.S., and Kerrich, R.
1979: Geochemistry and field relations of lode gold deposits in felsic igneous intrusions (abstract); in Geoscience Research Seminar, Abstracts, ed. E.G. Pye; Ontario Geological Survey, p. 8.

Kirkham, R.V. and Thorpe, R.I.
1973: Studies of gold-copper deposits suggest red metal may be used as guide to gold; The Northern Miner, November 29, p. 55, 57.

Latulippe, M.
1980: An overview of the geology of gold prospects and developments in N.W. Quebec; in The Canadian Institute of Mining and Metallurgy, Geology Division CIM Gold Symposium and Field Excursion, Val d'Or-Kirkland Lake-Timmins, p. 1-15.

Sinclair, W.D.
1982: Gold deposits of the Matachewan area, Ontario; in Geology of Canadian Gold Deposits, ed. R.W. Hodder and W. Petruk; Canadian Institute of Mining and Metallurgy, Special Volume 24, p. 83-93.

Thomson, J.E., Charlewood, G.H., Griffin, K., Hawley, J.E., Hopkins, H., MacIntosh, C.G., Ogrizlo, S.P., Perry, O.S., and Ward, W.
1948: Geology of the Main Ore Zone at Kirkland Lake; Ontario Department of Mines, v. 57, pt. 5, p. 54-188.

16 CARBONATITE-HOSTED DEPOSITS

Currie, K.L.
1976: The Alkaline Rocks of Canada; Geological Survey of Canada, Bulletin 239, 228 p.

Dawson, K.R.
1974: Niobium (Columbium) and Tantalum in Canada; Geological Survey of Canada, Economic Geology Report 29, 157 p.

Gold, D.P., Vallee, M., and Charlette, J.P.
1967: Economic geology and geophysics of the Oka Alkaline Complex, Quebec; Canadian Mining and Metallurgical Bulletin, v. 60, no. 666, p. 1131-1144.

Heinrich, E.W.
1966: The Geology of Carbonatites; Rand McNally & Company, 555 p.

Palabora Mining Company Limited Mine Geological and Mineralogical Staff
1976: The geology and the economic deposits of copper, iron and vermiculite in the Palabora Igneous Complex: a brief review; Economic Geology, v. 71, p. 177-192.

Parker, J.G. and Baroch, C.T.
1971: The rare-earth elements, yttrium, and thorium, with a chapter on Resources by J.W. Adams; United States Bureau of Mines Information Circular 8476, 92 p.

Sandvik, P.Q. and Erdosh, G.
1977: Geology of the Cargill phosphate deposit in northern Ontario; Canadian Mining and Metallurgical Bulletin, v. 69, no. 777, p. 90-96.

Vallée, M., and Dubuc, F.
1970: The St. Honoré carbonatite complex, Quebec; Canadian Mining and Metallurgical Bulletin, v. 63, no. 704, p. 1384-1394.

17 PORPHYRY COPPER, MOLYBDENUM, TUNGSTEN

Gustafson, L.B.
1978: Some major factors of porphyry copper genesis; Economic Geology, v. 73, p. 600-607.

Gustafson, L.B. and Hunt, J.P.
1975: The porphyry copper deposit at El Salvador, Chile; Economic Geology, v. 70, 857-912.

Lowell, J.D. and Guilbert, J.M.
1970: Lateral and vertical alteration-mineralization zoning in porphyry ore deposits; Economic Geology, v. 65, p. 373-408.

McMillan, W.J. and Panteleyev, P.
1980: Ore deposit models – 1. Porphyry copper deposits; Geoscience Canada, v. 7, p. 52-63.

Sutherland Brown, A. (editor)
1976: Porphyry deposits of the Canadian Cordillera; Canadian Institute of Mining and Metallurgy, Special Volume 15, 510 p.

Titley, S.R., editor
1982: Advances in geology of the porphyry copper deposits - southwestern North America; The University of Arizona Press, Tuscon, Arizona, 560 p.

Titley, S.R. and Beane, R.E.
1981: Porphyry copper deposits; in Economic Geology Seventy-Fifth Anniversary Volume, 1905-1980, ed. B.J. Skinner; Economic Geology Publishing Co., p. 214-269.

Wallace, S.R., Muncaster, N.K., Jonson, D.C., Mackenzie, W.B. Bookstrom, A.A. and Surface, V.A.
1968: Multiple intrusion and mineralization at Climax, Colorado; in Ore Deposits of the United States, 1933-1967 (Graton-Sales Volume), ed. J.D. Ridge; American Institute of Mining, Metallurgical, and Petroleum Engineers, Inc., New York, p. 605-640.

Westra, G. and Keith, S.B.
1981: Classification and genesis of stockwork molybdenum deposits; Economic Geology, v. 76, no. 4, p. 844-873.

White, W.H., Bookstrom, A.A., Kamilli, R.J., Ganster, M.W., Smith, R.P., Ranta, D.E., and Steininger, R.C.
1981: Character and origin of Climax-type molybdenum deposits; in Economic Geology Seventy-fifth anniversary volume, 1905-1980, ed. B.J. Skinner; Economic Geology Publishing Co., p. 270-316.

18 VEIN-STOCKWORK TIN

Mulligan, R.
1975: Geology of Canadian tin occurrences; Geological Survey of Canada, Economic Geology Report No. 28, 155 p.

Richardson, J.M.G., Spooner, E.T.C., and McAuslan, D.A.
1982: The East Kemptville tin deposit, Nova Scotia: an example of a large tonnage, low grade, greisen-hosted deposit in the endocontact zone of a granite batholith; in Current Research, Part B, Geological Survey of Canada, Paper 82-1B, p. 27-32.

Taylor, R.G.
1979: Geology of Tin Deposits; Elsevier Scientific Publishing Company, Amsterdam, 543 p.

19 SKARN DEPOSITS

19.1 Skarn Tungsten

Dawson, K.M. and Dick, L.A.
1978: Regional metallogeny of the Northern Cordillera: Tungsten and base metal-bearing skarns in southeastern Yukon and southwestern Mackenzie; in Current Research, Part A, Geological Survey of Canada, Paper 78-1A, p. 287-292.

Dick, L.A. and Hodgson, C.J.
1982: The MacTung W-Cu(Zn) contact metasomatic and related deposits of the Northeastern Canadian Cordillera; Economic Geology, v. 77, p. 845-867.

Einaudi, M.T., Meinert, L.D., and Newberry, R.J.
1981: Skarn deposits; in Economic Geology Seventy-fifty Anniversary Volume, 1905-1980, ed. B.J. Skinner; Economic Geology Publishing Co., p. 317-391.

John, Y.W.
1963: Geology and origin of Sangdong tungsten mine, Republic of Korea; Economic Geology, v. 58, p. 1285-1300.

Little, H.W.
1959: Tungsten deposits of Canada; Geological Survey of Canada, Economic Geology Report 17, p. 104-114.

Newberry, R.J.
1982: Tungsten-bearing skarns of the Sierra Nevada. I. The Pine Creek Mine, California; Economic Geology, v. 77, p. 823-844.

19.2 Skarn Zinc-Lead-Silver

Dawson, K.M. and Dick, L.A.
1978: Regional Metallogeny of the Northern Cordillera: Tungsten and base metal-bearing skarns in southeastern Yukon and southwestern Mackenzie; in Current Research, Part A, Geological Survey of Canada Paper 78-1A, p. 287-292.

Dick, L.A.
1980: A comparative study of the geology, mineralogy, and conditions of formation of contact metasomatic mineral deposits in the NE Canadian Cordillera; Ph.D. thesis, Queen's University, Kingston, 473 p.

Einaudi, M.T., Meinert, L.D., and Newberry, R.J.
1981: Skarn deposits; in Economic Geology, Seventy-fifth Anniversary Volume, 1905-1980, ed. B.J. Skinner; Economic Geology Publishing Co., p. 317-391.

Yun, S. and Einaudi, M.T.
1982: Zinc-lead skarns of the Yeonhwa-Ulchin District, South Korea; Economic Geology, v. 77, p. 1013-1032.

19.3 Skarn Iron

Einaudi, M.T., Minert, L.D., and Newberry, R.J.
1981: Skarn deposits; in Economic Geology Seventy-fifth Anniversary Volume, 1905-1980, ed. B.J. Skinner; Economic Geology Publishing Co., p. 317-391.

Eugster, H.P. and Chou, I-M.
1979: A model for the deposition of Cornwall-type magnetite deposits; Economic Geology, v. 74, p. 763-774.

Gross, G.A.
1965: Geology of Iron Deposits in Canada. Volume I. General Geology and Evaluation of Iron Deposits; Geological Survey of Canada, Economic Geology Report 22, 181 p.

1967: Geology of Iron Deposits in Canada. Volume II. Iron Deposits in the Appalachian and Grenville Regions of Canada; Geological Survey of Canada, Economic Geology Report 22, 111 p.

Morrison, G.W.
1980: Stratigraphic control of Cu-Fe skarn ore distribution and genesis at Craigmont, British Columbia; Canadian Institute of Mining and Metallurgy Bulletin, v. 73, no. 820, p. 109-123.

Rose, A.W.
1972: Favourability for Cornwall-type magnetite deposits in Pennsylvania using geological, geochemical and geophysical data in a discriminant function; Journal of Geochemical Exploration. v. I, p. 181-194.

United Nations, Department of Economic and Social Affairs
1970: Survey of World Iron Ore Resources; United Nations Publication, 479 p.

19.4 Skarn Copper

Allcock, J.B.
1982: Skarn and porphyry copper mineralization at Mines Gaspé, Murdochville, Quebec; Economic Geology, v. 77, p. 971-999.

Atkinson, W.W. and Einaudi, M.T.
1978: Skarn formation and mineralization in the contact aureole at Carr Fork, Bingham, Utah; Economic Geology, v. 73, p. 1326-1365.

Einaudi, M.T.
1982: General features and origin of skarns associated with porphyry copper plutons: Southwestern North America; in Advances in Geology of the Porphyry Copper Deposits - Southwestern North America, ed. S.R. Titley, p. 185-210.

Einaudi, M.T. and Burt, D.M.
1982: Introduction – terminology, classification, and composition of skarn deposits. A special issue devoted to skarn deposits; Economic Geology, v. 77, no. 4, p. 745-754.

Einaudi, M.T., Meinert, L.D., and Newberry, R.J.
1981: Skarn deposits; in Economic Geology Seventy-fifth Anniversary Volume, 1905-1980, ed. B.J. Skinner; Economic Geology Publishing Co., p. 317-391.

Morrison, G.W.
1980: Stratigraphic control of Cu-Fe skarn ore distribution and genesis at Craigmont, British Columbia; Canadian Institute of Mining and Metallurgy Bulletin, v. 73, no. 820, p. 109-123.

Shimazaki, H.
1980: Characteristics of skarn deposits and related acid magmatism in Japan; Economic Geology, v. 75, p. 173-183.

20 NEPHELINE- AND CORUNDUM-BEARING ALKALINE GNEISS

Allen, J.B. and Charsley, T.J.
1968: Nepheline syenite and phonolite; Institute of Geological Sciences, Mineral Resources Division, London, 169 p.

Currie, K.L.
1975: The alkaline rocks of Canada; Geological Survey of Canada, Bulletin 239, 228 p.

Hewitt, D.F.
1961: Nepheline syenite deposits of southern Ontario; Ontario Department of Mines Annual Report, v. 69, part 8, 194 p.

Sørenson, H., editor
1974: The alkaline rocks; John Wiley and Sons, New York, 622 p.

21 UNCONFORMITY-ASSOCIATED URANIUM

Economic Geology
1978: Special Issue Devoted to the Geology and Geochemistry of Uranium; v. 73, no. 8, p. 1401-1748.

Hoeve, J. and Sibbald, T.I.I.
1978: Mineralogy and geological settings of unconformity-type uranium deposits in northern Saskatchewan; in Uranium Deposits, Their Mineralogy and Origin, Mineralogical Association of Canada Short Course Handbook, Volume 3, ed. M.M. Kimberly; University of Toronto Press, Toronto, p. 457-474.

Tremblay, L.P.
1979: Comparaison de la géologie des gîtes d'uranium des types Beaverlodge et Discordance du nord du Saskatchewan; Université du Québec à Chicoutimi, Colloque sur la Prospective minérale du Québec, p. 191-196.

1982: Geology of the uranium deposits related to the sub-Athabasca Unconformity, Saskatchewan; Geological Survey of Canada, Paper 81-20, 56 p.

22 ARSENIDE VEIN SILVER, URANIUM

Mursky, G.
1973: Geology of the Port Radium Map-Area District of Mackenzie; Geological Survey of Canada, Memoir 374, 40 p.

Petruk, W. and Jambor, J.L. (compilers and principal authors)
1971: The silver-arsenide deposits of the Cobalt-Gowganda region, Ontario; The Canadian Mineralogist, v. 11, part 1, 429 p.

Robinson, B.W. and Ohmoto, H.
1973: Mineralogy, fluid inclusions, and stable isotopes of the Echo Bay U-Ni-Ag-Cu deposits, Northwest Territories, Canada; Economic Geology, v. 68, p. 635-656.

Withers, R.L.
1979: Mineral deposits of the Northrim Mine and a brief inquiry into the genesis of veins of the Ag, Bi, Ni, Co, As type; M.Sc. thesis, University of Alberta.

23 VEIN URANIUM

Beck, L.S.
1969: Uranium deposits of the Athabasca Region, Saskatchewan; Saskatchewan Department of Mineral Resources, Report 126, 139 p.

Lang, A.H., Griffith, J.W. and Steacy, H.R.
1962: Canadian deposits of Uranium and Thorium; Geological Survey of Canada, Economic Geology Report 16 (second edition), 324 p.

Ruzicka, V.
1971: Geological comparison between East European and Canadian uranium deposits; Geological Survey of Canada, Paper 70-48, 196 p.

Tremblay, L.P.
1972: Geology of the Beaverlodge mining area, Saskatchewan; Geological Survey of Canada, Memoir 367, 265 p.

1978: Geologic setting of the Beaverlodge-type of vein-uranium deposits and its comparison to that of the unconformity-type; in Uranium Deposits, Their Mineralogy and Origin, Mineralogical Association of Canada Short Course Handbook, Volume 3, ed. M.M. Kimberly; University of Toronto Press, Toronto, p. 431-456.

24 VEIN COPPER

Carr, J.M.
1971: Geology of the Churchill copper deposit; The Canadian Mining and Metallurgical Bulletin, v. 64, p. 50-54.

Duquette, G.
1970: Archean stratigraphy and ore relationships in the Chibougamau district; Quebec Department of Natural Resources, Special Paper 8, 16 p.

Hammer, D.F. and Peterson, D.W.
1968: Geology of the Magma mine area, Arizona; in Ore deposits of the United States, 1933-1967 (Graton-Sales Volume), ed. J.D. Ridge; American Institute of Mining, Metallurgical and Petroleum Engineers, Inc., New York, p. 1282-1310.

25 FELSIC INTRUSION-ASSOCIATED SILVER-LEAD-ZINC VEINS

Boyle, R.W.
1965: Geology, geochemistry, and origin of the lead-zinc-silver deposits of the Keno Hill-Galena Hill area, Yukon Territory; Geological Survey of Canada, Bulletin 111, 302 p.

Fyles, J.T.
1967: Geology of the Ainsworth-Kaslo area, British Columbia; British Columbia Department of Mines and Petroleum Resources, Bulletin 53, 125 p.

Little, H.W.
1960: Nelson map-area, west half, British Columbia; Geological Survey of Canada, Memoir 308, 305 p.

26 ULTRAMAFIC-HOSTED ASBESTOS

Anhaeusser, C.R.
1976: The nature of chrysotile asbestos occurrences in southern Africa: A review; Economic Geology, v. 71, p. 96-116.

Laurent, R. and Hébert, Y.
1979: Paragenesis of serpentine assemblages in harzburgite and dunite cumulate from the Quebec Appalachians; Canadian Mineralogist, v. 17, p. 857-869.

Riordan, P.H.
1975: Geology of asbestos deposits of southeastern Quebec; Ministère des Richesses Naturelles du Québec, Étude Speciale 18.

Riordan, P.H.
1981: Geology of asbestos deposits; Society of Mining Engineers, American Institute of Mining Engineers, Littleton, Colorado, 118 p.

APPENDIX 1

LOCATIONS OF CANADIAN MINERAL DEPOSITS CITED IN 'SUMMARIES' SECTION

Name	Prov.	NTS	Lat.	Long.	Deposit Type
Adanac	B.C.	104 N 11	59°45'	133°25'	17.c
Advocate Mine	Nfld.	12 H 1	50°00'	56°10'	26
Afton	B.C.	92 I 10	50°40'	120°30'	17.b
Ainsworth District	B.C.	82 F 14,15	49°45'	117°00'	25
Albert Formation	N.B.	southern New Brunswick			1.b
Alice Arm	B.C.	103 P 6	55°25'	129°25'	17.c
Asbestos Hill	Que.	35 H 13	61°50'	74°00'	26
Athabasca Basin	Sask.	northern Saskatchewan			4.b
Atikokan	Ont.	52 B 14	48°45'	91°25'	3
Barnat Mine	Que.	32 D 1	48°10'	78°05'	15.b
Beaverlodge area	Sask.	74 N 7,8, 9,10	59°30'	108°30'	23
Bell-King-Beaver Mines	Que.	21 L 3	46°05'	71°20'	26
Belmoral	Que.	32 C 4	48°10'	77°40'	15.a
Betts Cove	Nfld.	2 E 13	49°50'	55°50'	9.1.a
Big Trout Lake	Ont.	53 H 12	53°40'	89°30'	13.1
Bigwood	Ont.	41 I 2 H 15	46°00'	80°40'	20.a
Birchtree Mine	Man.	63 P 12	55°40'	97°55'	12.1.b
Bird River Sill	Man.	52 L 15,16	95°30'	50°30'	13.1
Black Lake	Que.	21 L 3	46°00'	71°20'	26
Blizzard deposit	B.C.	82 E 10	49°40'	118°55'	6.4
Blue Mountain	Ont.	31 C 12	44°40'	77°55'	20.a
Bluff Head Mine	Nfld.	12 B 15	48°45'	58°35'	13.2
Bralorne	B.C.	92 J 15	50°45'	122°50'	11
Brenda	B.C.	92 H 16	49°55'	120°00'	17.a
British Canadian Mine	Que.	21 L 3	46°00'	71°20'	26
Bruce Mines area	Ont.	41 J 15	46°20'	83°45'	24
Brunswick No. 12 deposit	N.B.	21 P 5	47°30'	65°55'	9.1.b
Buchans	Nfld.	12 A 15	48°50'	56°50'	9.1.b
Burgess Mine	Ont.	31 F 5	45°15'	77°40'	20.b
Camflo Mine	Que.	32 D 1	48°10'	78°00'	15.a
Campbell Red Lake Mine	Ont.	52 N 4	51°05'	93°45'	11
Cantung	N.W.T.	105 H 16	61°55'	128°15'	19.1
Cargill Twp. deposit	Ont.	42 G 8	49°20'	82°50'	16.a
Cariboo District	B.C.	93 A 13,14 H 3,4	53°00'	121°30'	5.2
Caribou Mine	Que.	21 L 3	46°00'	71°20'	13.2
Carolin	B.C.	92 H 11	49°30'	121°20'	11
Cassiar Asbestos Mine	B.C.	104 P 5	59°20'	129°50'	26
Cassiar Batholith	Y.T.	105 B 104 O	60°00'	131°00'	19.2
Cathy deposit	Y.T.	105 O 7	63°15'	130°35'	9.3
Chaudière area	Que.	21 D 15 E 12	46°10'	70°40'	5.2
Chibougamau District	Que.	32 G 16	49°00'	74°30'	24
Chrome Lake-Puddy Lake	Ont.	52 H 13	50°00'	89°30'	13.1
Chromeraine (Reed-Belanger) Mine	Que.	21 L 3	46°00'	71°20'	13.2
Churchill area	B.C.	94 K	58°30'	125°30'	24
Cirque Mine	B.C.	104 N 14	59°46'	133°15'	9.2
Clinton Creek Mine	Y.T.	116 C 7	64°25'	140°45'	26
Cluff Lake "D" orebody	Sask.	74 K 5	58°20'	109°30'	21
Coast Copper Mine	B.C.	92 L 6	50°25'	127°15'	19.4
Cobalt District	Ont.	31 M 5	47°25'	79°40'	22
Con Mine	N.W.T.	85 J 8	62°25'	114°20'	11
Contwoyto Lake	N.W.T.	76 E 14	65°45'	111°15'	7.b
Copper Mountain	B.C.	92 H 7	49°20'	120°35'	17.b
Coppermine River area	N.W.T.	86 N,O	67°30'	114°00'	10
Craigmont Mine	B.C.	92 I 2	50°10'	120°55'	19.4
Craigmont Mine	Ont.	31 F 5	45°20'	77°35'	20.b
Crystal Lake Gabbro	Ont.	52 A 4	48°05'	89°40'	13.1
Davis Hill Quarry	Ont.	31 F 4	45°05'	77°45'	20.a
Detour Lake	Ont.	32 L 4	50°00'	79°40'	7.d

Note: Latitude and longitude given are rounded to the nearest 05'. For less precisely defined locations (eg. Beaverlodge area, Cassiar batholith) the centre is approximately located.

Name	Prov.	NTS	Lat.	Long.	Deposit Type
Dome Mine	Ont.	42 A 6	48°30'	81°15'	11
Dorchester	N.B.	21 H 16	45°55'	64°45'	6.3.b
Doyon-Silver Stack deposit	Que.	32 D 7	48°15'	78°30'	7.c
Dumagami deposit	Que.	32 D 8	48°15'	78°25'	7.c
Dumont Sill	Que.	32 D 9	48°40'	78°25'	12.1.b
Eagle Gold Mine	Que.	32 E 8	49°30'	78°20'	7.c
East Kemptville	N.S.	21 A 4	44°05'	65°40'	18
Echo Bay Mine	N.W.T.	86 K 4	60°05'	118°00'	22
Elliot Lake	Ont.	41 J 7	46°25'	82°40'	5.1.a
Elsa	Y.T.	105 M 14	63°55'	135°30'	25
Endako	B.C.	93 K 3	54°05'	125°05'	17.c
Faro	Y.T.	105 K 6	62°21'	133°20'	9.2
Fay-Verna Mine	Sask.	74 N 9	59°35'	108°30'	23
Fernie Synclinorium	Alta/B.C.	82 G,J	southern Alta./B.C.		4.a
Flin Flon Mine	Man.-Sask.	63 K 13	54°45'	101°55'	9.1.a
French Mine	Que./Lab.	23 J 15	54°50'	66°55'	3
Frenchman's Cap	B.C.	82 M 8	51°20'	118°30'	20.a
Galena Hill District	Y.T.	105 M 14	63°55'	135°25'	25
Galore Creek	B.C.	104 G 3,4	57°05'	131°30'	17.b
Gaspé Copper Mine	Que.	22 A 13	49°00'	65°30'	19.4
George Lake	Sask.	64 E 5	57°30'	103°45'	6.2
Geraldton District	Ont.	42 E	49°45'	87°00'	7.a
Giant Mascot	B.C.	92 H 5	49°30'	121°30'	12.2.c
Giant Yellowknife Mine	N.W.T.	85 J 8	62°30'	114°20'	11
Granisle	B.C.	93 L 16	54°55'	126°10'	17.a
Great Bear Lake District	N.W.T.	86 E 13 L 1	66°00'	118°00'	22
Great Lakes Nickel	Ont.	52 A 4	44°05'	89°35'	12.2.a
Griffith Mine	Ont.	52 K 14	50°50'	93°25'	2.3
Gunnar Mine	Sask.	74 N 7	59°25'	108°50'	23
Gypsumville	Man.	65 O 15	51°45'	98°35'	1.a
Haliburton-Bancroft-Renfrew belt	Ont.	31 C,D,E,F	eastern Ontario		20.a
Hector-Calumet	Y.T.	105 M 14	63°55'	135°25'	25
Helen Mine	Ont.	42 C 2	48°00'	84°45'	2.3
Highland Bell	B.C.	82 E 6	49°25'	119°05'	25
Highland Valley	B.C.	92 I	50°30'	121°00'	17.a
Hollinger Mine	Ont.	42 A 6	48°30'	81°20'	15.a
Horne Mine	Que.	32 D 6	48°15'	79°00'	9.1.a
Howards Pass	Y.T.	105 I 6	62°30'	129°10'	9.2
Howey-Hasaga Mine	Ont.	52 N 4	51°00'	93°50'	15.c
HPH deposit	B.C.	92 L 12	50°40'	127°50'	19.2
Icon area	Que.	32 I 4	50°15'	73°50'	24
Ingerbelle Mine	B.C.	92 H 7	49°20'	120°35'	19.4
Island Copper	B.C.	92 L 11	50°40'	127°30'	17.a
Ivry Mine	Que.	31 J 1	46°05'	74°20'	14.a
Jason deposit	Y.T.	105 O 1	63°10'	130°15'	9.2
Jeffrey Mine	Que.	21 E 13	45°45'	71°44'	26
Keno Hill District	Y.T.	105 M 14	63°55'	135°15'	25
Kerr Addison Mine	Ont.	32 D 4	48°10'	79°35'	11
Key Lake	Sask.	74 H 4	57°10'	105°40'	21
Kidd Creek Mine	Ont.	42 A 11	48°40'	81°20'	9.1.a
Kirkland Lake area	Ont.	42 A 1 32 D 4	48°00'	80°00'	15.b
Klondike area	Y.T.	115 O 15 116 B 2,3	63°45'	139°00'	5.2
Knob Lake	Que./Lab.	23 J 15	54°45'	66°20'	2.2
Lac Allard	Que.	12 L 5,11	50°30'	63°30'	14.a
Lac Doré Complex	Que.	32 G 16	49°55'	74°10'	14.b
Lac des Isles	Ont.	52 H 4	49°10'	89°35'	12.2.b
Lac des Montagnes	Que.	32 O 12	51°40'	75°55'	13.1
Lackner Lake	Ont.	41 O 14	47°50'	83°15'	16.a
Lake Harbour, Baffin Is.	N.W.T.	25 N 4	63°10'	69°40'	20.a
Lake St. Joseph	Ont.	52 O 1	51°00'	90°30'	2.3
Lamaque Mine	Que.	32 C 4	48°05'	77°45'	15.a
Langmuir Mine	Ont.	42 A 6	48°20'	81°00'	12.1.a
Logtung	Y.T.	105 B 4	60°00'	131°35'	17.c
Lucky Jim	B.C.	82 K 3	50°00'	117°10'	25
Lupin Mine	N.W.T.	76 E 14	65°45'	111'15'	7.b
Lynn Lake	Man.	64 C 14	56°50'	101°05'	12.2.c
Mactung	Y.T.	105 O 8	63°15'	130°10'	19.1
Madelaine Mine	Que.	22 G 1 B 16	49°00'	60°00'	19.4

Name	Prov.	NTS	Lat.	Long.	Deposit Type
Magpie Mountain	Que.	22 P 8	51°25'	64°05'	14.b
Mamainse Point	Ont.	41 N 2	47°05'	84°45'	10
Manibridge Mine	Man.	63 J 10	54°40'	98°50'	12.1.b
Marbridge Mine	Que.	32 D 8	48°20'	78°10'	12.1.a
Marmora	Ont.	31 C 5	44°30'	77°40'	19.3.b
Martison Lake	Ont.	42 J 6	50°25'	83°10'	16.a
McIntyre Mine	Ont.	See "Pamour Schumacher mine"			
Meat Cove	N.S.	11 N 2	47°00'	60°35'	19.2
		K 15			
Meguma Group	N.S.	11 D,E,F	northeastern N.S.		8.2
Millenbach Mine	Que.	32 D 6	48°20'	79°05'	9.1.a
Montauban	Que.	31 I 16	46°50'	72°20'	7.c
Monteagle deposit	Ont.	31 F 4	45°10'	77°50'	20.a
Montreal Mine	Que.	21 E 14	46°00'	71°15'	13.2
Mount Billings Batholith	Y.T.	105 A,B,G,H	61°00'	129°00'	19.2
Mount Copeland	B.C.	82 M 1	51°10'	118°30'	20.c
Mount Pleasant	N.B.	21 G 7	45°25'	66°50'	17.c
Mount Pleasant	N.B.	21 G 7	45°25'	66°50'	18
Mount Wright	Que./Lab.	23 B 11,14	52°45'	67°10'	2.2
Mountain Lake	N.W.T.	86 N 7	67°20'	116°55'	6.4
Munro Mine	Ont.	42 A 9	48°35'	80°15'	26
Muskox Intrusion	N.W.T.	86 J 11,14	67°00'	115°10'	13.1
		O 13			
Nemegosenda Lake	Ont.	42 B 3	48°00'	83°05'	16.a
New Denver District	B.C.	82 F 14	50°00'	117°15'	25
		K 3			
Newboro Lake	Ont.	31 C 9	44°40'	76°20'	14.b
Newfoundland Zinc	Nfld.	12 I 6	50°20'	57°30'	6.1
Oka	Que.	31 G 8,9	45°30'	74°00'	16.a
Opemiska District	Que.	32 G 15	49°45'	74°50'	24
Oro deposit	Y.T.	105 I 12	62°40'	129°50'	9.3
Otto Fiord Formation	N.W.T.	Ellesmere and Sverdrup islands			1.a
Padlei	N.W.T.	65 H 15	61°55'	96°30'	5.1.a
Pamour No. 1 Mine	Ont.	42 A 6	48°30'	81°10'	11
Pamour Schumacher Mine	Ont.	42 A 6	48°30'	81°15'	15.a
Peace River (Clear Hills Ironstone)	Alta.	84 D 11	56°30'	119°00'	2.1
Phoenix Mine	B.C.	82 E 2	49°05'	118°35'	19.4
Pickle Crow District	Ont.	86 O 5	67°20'	115°50'	7.a
Pine Point	N.W.T.	85 B 16	60°50'	114°25'	6.1
Pipe Mine	Man.	63 O 8	55°30'	98°10'	12.1.b
Polaris Mine	N.W.T.	68 H 8	75°25'	96°55'	6.1
Prairie Formation	Sask.	southern Saskatchewan			1.a
Princess Sodalite Mine	Ont.	31 F 4	45°05'	77°50'	20.a
Puddy Lake - Chrome Lake	Ont.	52 H 13	50°00'	89°30'	13.1
Rabbit Lake	Sask.	64 L 4	58°10'	103°45'	21
Rapid Creek - Big Fish River area	Y.T.	117 A	northern Yukon		4.a
Rayrock Mine	N.W.T.	85 N 7	63°25'	116°30'	23
Redstone deposit	N.W.T.	95 L 10	62°40'	126°35'	6.3.a
Reed-Belanger (Chromeraine) Mine	Que.	21 L 3	46°00'	71°20'	13.2
Ruth Mine	Que./Lab.	23 J 15	54°50'	66°50'	3
Saint Charles	Que.	22 D 11	48°30'	71°30'	14.b
Sakami Lake	Que.	33 F 2	53°10'	76°55'	5.1.a
Salina Formation	Ont.	40 J	southwestern Ontario		1.a
Salmo District	B.C.	82 F 3,6	49°15'	117°15'	19.1
San Antonio Mine	Man.	52 M 4	51°00'	95°40'	15.c
Scottie Creek	B.C.	92 I 14	51°00'	121°25'	13.2
Shebandowan	Ont.	52 B 9	48°35'	90°15'	12.1.c
Sherman Mine	Ont.	31 M 4	47°05'	79°50'	2.3
Slocan District	B.C.	82 F 11,14	49°45'	117°25'	25
St. Honoré	Que.	22 D 11	48°30'	71°10'	16.a
St. Stephen	N.B.	21 G 3	45°15'	67°20'	12.2.c
St. Urbain	Que.	21 M 8	47°30'	70°35'	14.a
Sterrett Mine	Que.	31 H 9	45°40'	72°00'	13.2
Sudbury	Ont.	41 I 6	46°30'	81°15'	12.2.a
Sullivan Mine	B.C.	82 F 9	49°40'	116°00'	9.2
Sulphur Creek	B.C.	94 N 4	59°05'	125°40'	9.3

Name	Prov.	NTS	Lat.	Long.	Deposit Type
Sustut deposit	B.C.	94 D 10	56°35°	126°40'	10
Tasu	B.C.	103 C 9,16	52°45'	132°00'	19.3.a
Tea deposit	Y.T.	105 O 2	63°00'	130°35'	9.3
Terra Mine	N.W.T.	86 E 9	65°35'	118°05'	22
Texada Island	B.C.	92 F 9,10	49°40'	124°30'	19.3
Thetford Mines	Que.	21 L 3	46°05'	71°20'	26
Timmins Mine	Que./Lab.	23 J/14	54°55'	67°05'	3
Tom deposit	Y.T.	105 O 1	63°10'	130°10'	9.2
Trident Mountain	B.C.	82 M 16	51°55'	118°10'	20.a
Ungava deposits	Que.	35 H 11,12	61°40'	73°30'	12.1.c
Wabana Mine	Nfld.	1 N 10	47°40'	53°00'	2.1
Wabush Lake	Que./Lab.	23 G 2	53°00'	67°00'	2.2
Walton	N.S.	21 H 1	45°10'	64°00'	9.2
Western Mines	B.C.	92 F 12	49°35'	125°35'	9.1.b
White River	Y.T.	115 F 15	61°45'	140°45'	10
Whitehorse Copper Mine	Y.T.	105 D 11	60°40'	135°05'	19.4
Wilmar Mine	Ont.	52 N 4	51°00'	93°45'	15.a
Windermere	B.C.	82 J 5	50°25'	115°55'	1.a
Windsor Group	N.S.		central Nova Scotia		1.a
Woodstock	N.B.	21 J 4	46°10'	67°40'	2.3
Yava	N.S.	11 F 15,16	45°50'	60°25'	6.2
Yellowknife Supergroup	N.W.T.		Slave province, central N.W.T.		8.2
York Harbour	Nfld.	12 G 1	49°00'	58°20'	9.1.a
Young-Davidson Mine	Ont.	41 P 15	47°55'	80°40'	15.b
Zip deposit	B.C.	92 L 7	50°20'	126°55'	19.2

NOTES

NOTES

NOTES

NOTES

NOTES

NOTES